# Fuzzy Clustering for Responsive Data

ARASH ABADPOUR

Arash Abadpour (arash@abadpour.com) is with the Imaging Group, Epson Edge, Epson Canada Limited.

The author acknowledges that this text has been submitted for publication and that this file will be deleted upon publication.

# Contents

**1 Introduction**      **1**

**2 Literature Review**      **3**

    2.1    Notion of Membership . . . . . . . . . . . . . . . . . . . . . . . . . . . . . 3

    2.2    Prototype-based Clustering . . . . . . . . . . . . . . . . . . . . . . . . . 5

    2.3    Robustification . . . . . . . . . . . . . . . . . . . . . . . . . . . . . . . . 7

    2.4    Number of Clusters . . . . . . . . . . . . . . . . . . . . . . . . . . . . . . 9

    2.5    Weber Problem . . . . . . . . . . . . . . . . . . . . . . . . . . . . . . . . 10

**3 Developed Method**      **13**

    3.1    Model Preliminaries . . . . . . . . . . . . . . . . . . . . . . . . . . . . . 13

    3.2    Assessment of Loss . . . . . . . . . . . . . . . . . . . . . . . . . . . . . . 15

    3.3    Fuzzification . . . . . . . . . . . . . . . . . . . . . . . . . . . . . . . . . 16

    3.4    Problem Development . . . . . . . . . . . . . . . . . . . . . . . . . . . . 17

    3.5    Solution Strategy . . . . . . . . . . . . . . . . . . . . . . . . . . . . . . 18

    3.6    Classification & Outlier Removal . . . . . . . . . . . . . . . . . . . . . . 19

    3.7    Data Responsiveness . . . . . . . . . . . . . . . . . . . . . . . . . . . . 19

    3.8    Determination of $U$ and $\lambda$ . . . . . . . . . . . . . . . . . . . . . . . . 21

    3.9    Convergence . . . . . . . . . . . . . . . . . . . . . . . . . . . . . . . . . 22

    3.10   Implementation Notes . . . . . . . . . . . . . . . . . . . . . . . . . . . . 23

**4 Experimental Results**      **25**

    4.1    Models . . . . . . . . . . . . . . . . . . . . . . . . . . . . . . . . . . . . 26

        4.1.1    Grayscale Image Multi-Level Thresholding . . . . . . . . . . . . 27

        4.1.2    2D Euclidean Clustering . . . . . . . . . . . . . . . . . . . . . . 27

4.1.3   Plane Finding in Range Data . . . . . . . . . . . . . . . . . . . . . . . . . 28

4.2  Comparative Results . . . . . . . . . . . . . . . . . . . . . . . . . . . . . . . . . . . 28

    4.2.1   Grayscale Image Multi-Level Thresholding . . . . . . . . . . . . . . . . . . 29

    4.2.2   2D Euclidean Clustering . . . . . . . . . . . . . . . . . . . . . . . . . . . . . 32

    4.2.3   Plane Finding in Range Data . . . . . . . . . . . . . . . . . . . . . . . . . 35

4.3  Responsiveness . . . . . . . . . . . . . . . . . . . . . . . . . . . . . . . . . . . . . . . 37

    4.3.1   Grayscale Image Multi-Level Thresholding . . . . . . . . . . . . . . . . . . 37

    4.3.2   2D Euclidean Clustering . . . . . . . . . . . . . . . . . . . . . . . . . . . . . 38

    4.3.3   Plane Finding in Range Data . . . . . . . . . . . . . . . . . . . . . . . . . 39

4.4  Convergence . . . . . . . . . . . . . . . . . . . . . . . . . . . . . . . . . . . . . . . . . 40

**5  Conclusions**      **43**

# List of Figures

3.1   The process of determining the values of $\lambda$ and $U$. (a) $\lambda$. (b) $U$.   . . . . . . . . . . .   21

4.1   Standard image *House*. . . . . . . . . . . . . . . . . . . . . . . . . . . . . .   29

4.2   Results of applying FCM and the developed method for grayscale image multi-level thresholding on the Standard image *House*, shown in Figure 4.1, for $C = 2$. (a) FCM. (b) Proposed Method.   . . . . . . . . . . . . . . . . . . . . . . . . . . .   30

4.3   Results of applying FCM and the developed method for grayscale image multi-level thresholding on the Standard image *House*, shown in Figure 4.1, for $C = 3$ and $C = 4$. (a) FCM, $C = 3$. (b) Proposed method, $C = 3$. (c) FCM, $C = 4$. (d) Proposed method, $C = 4$.   . . . . . . . . . . . . . . . . . . . . . . . . . . .   32

4.4   Results of applying FCM and the developed method for 2D Euclidean clustering. Size of the data items denote their weight and their shade of gray indicates the probability of being an inlier. (a) FCM. (b) Proposed method. (c) Data items after they have responded to the clusters.   . . . . . . . . . . . . . . . . . . . . . .   33

4.5   Results of applying FCM and the developed method for 2D Euclidean clustering. (a) FCM. (b) Proposed method. (c) Data items after they have responded to the clusters.   34

4.6   Results of applying FCM and the developed method for plane finding in range data. (a) FCM. (b) Proposed method.   . . . . . . . . . . . . . . . . . . . . . . . .   35

4.7   Results of applying FCM and the developed method for plane finding in range data. (a) FCM. (b) Proposed method.   . . . . . . . . . . . . . . . . . . . . . . . .   36

4.8   Data responsiveness for grayscale image multi-level thresholding corresponding to the results shown in Figures 4.2 and Figure 4.3. . . . . . . . . . . . . . . . . . . .   38

4.9   Examination of the responses of the data items to the converged clusters for the result shown in Figure 4.6. (a) Before. (b) After. . . . . . . . . . . . . . . . . . .   39

# Abstract

The unsupervised clustering of a weighted group of data items, into sets which comply with a notion of homogeneity specific to a particular problem class, is a classical pattern classification problem. Among the different approaches to this problem, fuzzy and possibilistic clustering, and their different variations and combinations, have been studied in the last decades. That study, however, has often been focused on a particular data item or cluster model and has been generally geared towards identifying a particular model of homogeneity. Also, many available models are in essence engineered based on the intuition of the researchers and convoluted and complicated control parameters, regularization terms, and concepts are often incorporated into the mathematical models in order to yield acceptable results. In this paper, we advocate for a derivation-based approach, which utilizes Bayesian inference in order to assess the loss in a generic clustering problem, independent of the particularities of data and cluster models and the notion of homogeneity applicable to any particular problem class. Subsequently, we utilize the organic framework developed in this work in order to address data items which actively respond to the clustering effort. This is distinctly different from the available literature, in which data items are passively subjected to the clustering process. The utilization of an inference-based loss model, which avoids exogenous provisions based on intuition and researcher heuristics, as well as independence from a particular problem class and the introduction of the concept of data responsiveness, to our understanding, are novel to this paper.

# Chapter 1

# Introduction

Unsupervised clustering of a group of data items into homogenous sets, is a primitive operation required in many different applications within fields such as signal and image processing. Constructing a class-independent clustering algorithm, however, is a non-trivial task because the data involved in these operations relates to physical phenomena with different characteristics. For example, one may discuss the proper clustering of vectors or other mathematical objects which represent measured entities governed by particular homogeneity models which are intrinsic to the underlying physical phenomenon. In this paper, a *Problem Class* is the category of all clustering problems which involve the same data and cluster models. Nevertheless, while different problem classes utilize data items of different types and a varied range of notions of homogeneity, they all share a common structure, which is the dissection of a set of data items into a number of homogenous sets and, potentially, a set of outliers. Hence, rather than designing a separate clustering algorithm per problem class, it is distinctly advantageous to abstract out the mathematical models of the data and the clusters and to discuss data clustering in generic terms. Note that in this framework, a *Problem Instance* is one particular instance which belongs to a problem class, as defined above.

A direct outcome of this abstraction is saved time and efforts when the need for data clustering in a new context arises. Additionally, the abstraction of the mathematical models allows for following a systematic approach and avoiding problem specific provisions. Such harmful activities include the use of engineered terms which intuitively and based on heuristics are suggested to yield "proper" clustering results and the utilization of exogenous parameters and constraints which one may suggest "ought to be included in the clustering process for a particular problem class". In this paper, we argue that the mathematical structure of the clustering problem must be organically derived based on a cost assessment process which is independent of the properties of any particular problem class.

Additionally, in this paper, we propose the concept of data responsiveness. In effect, we argue that the contemporary approaches to data clustering consider data items which offer themselves for clustering, but do not respond to the outcomes of the process. In those approaches, after the clustering effort converges, a Maximum Likelihood-style operation is utilized in order to separate the data into hard clusters. In this paper, however, we advise that the clustering process must be considered from the perspective of the data items as well and that the data items must be allowed to actively respond to the outcomes of the clustering process.

In this paper, we construct a generic data clustering problem and utilize Bayesian inference in order to assess the amount of loss when a particular clustering solution is considered. This model utilizes a robust loss function and employs concepts developed in the fuzzy clustering literature. We exhibit that the commonly utilized *Alternating Optimization* framework can be replaced by an iterative process which eliminates the need for the successive calculation of membership and possibilistic terms, thus reducing computational load. We demonstrate that the separation of data into the different clusters as well as the determination of the potential outliers can be carried out without the explicit calculation of the membership terms. Then, we address data responsiveness, as discussed above.

The rest of this paper is organized as follows. First, in Chapter 2, we review the related literature and then, in Chapter 3, we present the developed method. Subsequently, in Chapter 4, we provide experimental results produced by the developed method and, in Chapter 5, we carry the concluding remarks.

# Chapter 2

# Literature Review

## 2.1 Notion of Membership

The notion of membership is a key point of distinction between different clustering algorithms. Essentially, membership may be *Hard* or *Fuzzy*. Within the context of hard membership, each data item belongs to one cluster and is different from all other clusters. The fuzzy membership regime, however, maintains that each data item in fact belongs to all clusters, with the stipulation that the degree of membership to different clusters is different. K-means and Hard C-means (HCM) [1, 2] clustering algorithms, as well as Iterative Self-Organizing Data Clustering (ISODATA) [3] algorithm, utilize hard membership values. The reader is referred to [4] and the references therein for a history of K-means clustering and other methods closely related to it.

With the introduction of Fuzzy Theory [5], many researchers incorporated this more natural notion into clustering algorithms [6, 7]. The premise for employing a fuzzy clustering algorithm is that fuzzy membership is more applicable in practical settings, where generally no distinct line of separation between clusters is present [8]. Additionally, from a practical perspective, it is observed that hard clustering techniques are extremely more prone to falling into local minima [1]. The reader is referred to [9, 10] for the wide array of fuzzy clustering methods developed in the past few decades.

Initial work on fuzzy clustering was done by Ruspini [11] and Dunn [12] and it was then generalized by Bezdek [9] into Fuzzy C-means (FCM). In FCM, data items, which are denoted as $x_1, \cdots, x_N$, belong to $\mathbb{R}^k$ and clusters, which are identified as $\psi_1, \cdots, \psi_C$, are represented as points in $\mathbb{R}^k$. FCM makes the assumption that the number of clusters, $C$, is known through a separate process or expert opinion and minimizes the following cost function,

$$\Delta = \sum_{c=1}^{C} \sum_{n=1}^{N} f_{nc}^{m} \|x_n - \psi_c\|^2 . \tag{2.1}$$

This objective function is heuristically suggested to result in appropriate clustering results and is constrained by,

$$\sum_{c=1}^{C} f_{nc} = 1, \forall n. \tag{2.2}$$

Here, $f_{nc} \in [0,1]$ denotes the membership of $x_n$ to $\psi_c$.

In (2.1), $m > 1$ is the *fuzzifier* (also called *weighing exponent* and *fuzziness*). The optimal choice for the value of the fuzzifier is a debated matter [13] and is suggested to be "an open question" [14]. Bezdek [15] suggests that $1 < m < 5$ is a proper range and utilizes $m = 2$. The use of $m = 2$ is suggested by Dunn [12] in his early work on the topic as well and also by Frigui et al. [16], among others [17]. Bezdek [18] provided physical evidence for the choice of $m = 2$ and Pal et al. [19] suggested that the best choice for $m$ is probably in the interval $[1.5, 2.5]$. Yu et al. [14] argue that the choices for the value of $m$ are mainly empirical and lack a theoretical basis. They worked on providing such a basis and suggested that "a proper $m$ depends on the data set itself" [14]. Nevertheless, it is known that larger values of $m$ soften the boundary between the clusters [20].

Recently, Zhou et al. [21] proposed a method for determining the optimal value of $m$ in the context of FCM. They employed four Cluster Validity Index (CVI) models and utilized repeated clustering for $m \in [1.1, 5]$ on four synthetic data sets as well as four real data sets adopted from the UCI Machine Learning Repository [22] (refer to [23] for a review of CVIs and [24] for coverage in the context of relational clustering). The range for $m$ in that work is based on previous research [25] which provided lower and upper bounds on $m$. The investigation carried in [21] yields that $m = 2.5$ and $m = 3$ are optimal in many cases and that $m = 2$ may in fact not be appropriate for an arbitrary set of data items. This result is in line with other works which demonstrate that larger values of $m$ provide more robustness against noise and the outliers. Nevertheless, significantly large values of $m$ are known to push the convergence towards the sample mean, in the context of Euclidean clustering [14]. Wu [26] analyzes FCM and some of its variants in the context of robustness and recommends $m = 4$. In [27], the authors address the related problem of fuzzy model construction. They set up the framework using $m = 2$ and proceed to find the optimal value of $m$ for different problem instances while maintaining $m \in [1.1, 5]$. They show that different values of $m$ are optimal within the context of different problem classes.

Rousseeuw et al. [28] suggested to replace $f_{nc}^m$ with $\alpha f_{nc} + (1 - \alpha)f_{nc}^2$, for a known $0 < \alpha < 1$. Klawonn et al. [29, 30] suggested to generalize this effort and to replace $f_{nc}^m$ with an increasing and differentiable function $g(f_{nc})$.

Pedrycz [31, 32, 33] suggested to modify (2.2) in favor of customized $\sum f_{nc}$ constraints for different values of $n$. That technique allows for the inclusion of *a priori* information into the clustering framework and is addressed as Conditional Fuzzy C-means (CFCM). The same modification is carried out in Credibilistic Fuzzy C-Means (CFCM) [34, 35], in which the "credibility" of data items is defined based on the distances between data items and clusters. Therefore, in that approach, (2.2) is modified in order to deflate the membership of outliers to the set of clusters (also see [36]). Customization of (2.2) is also carried out in Cluster Size Insensitive FCM (csiFCM) [37] in order to moderate the impact of data items in larger clusters on an smaller adjacent cluster. Leski [13] provides a generalized version of this approach in which $\sum \beta f_{nc}^\alpha$ is constrained.

Note that, many fuzzy and possibilistic clustering algorithms make the assumption that the data items are equally important. Weighted fuzzy clustering, however, works on data items which have an associated positive weight [38]. This notion can be considered as a marginal case of clustering fuzzy data [39]. Examples for this setting include clustering of a weighted set, clustering of sampled data, clustering in the presence of multiple classes of data items with different priorities [40], and a measure used in order to speed up the execution through data reduction [41, 42, 43, 44]. Nock et al. [45] formalize the case in which weights are manipulated in order to move the clustering results towards data items which are harder to include regularly. Chen et al. [46] utilize density motivated weights in order to reduce the impact of outliers (refer to [47] for different variants of this framework). Semi Supervised FCM (ssFCM) [48] uses weight factors based on an Euclidean norm in order to balance the sizes of different hyper-spherical shaped clusters based on user intervention. Note that the extension of FCM on weighted sets has been developed under different names, including Density-Weighted FCM (WFCM) [43], Fuzzy Weighted C-means (FWCM) [49], and New Weighted FCM (NW-FCM) [50].

## 2.2 Prototype-based Clustering

It is a common assumption that homogeneity depends on the distance between data items. This assumption is made implicitly when clusters are modeled as *prototypical* data items, also called *clustroids* or cluster *centroids* [51]. For example, the potential function approach considers data items as energy sources scattered in a multi-dimensional space and seeks peak values in the field [52] (also see [53]). We argue, however, that the *distance* between data items may not be meaningful and that what the clustering algorithm is to accomplish is the minimization of the distances between the data items and the clusters. For example, when the data is to be clustered into certain lower-dimensional subspaces, as is the case with Fuzzy C-Varieties (FCV) [54], the Euclidean distance between the data items is irrelevant. This is in contrast, for example, with the core assumption of the non-iterative graph-theory based approach Single Linkage (SL) [55], which constructs the clusters through traversing a graph in which the nodes represent the data items and the edges correspond to their pairwise distances.

Nevertheless, prototype-based clustering does not necessarily require explicitly present prototypes. For example, in kernel-based clustering, it is assumed that a non-Euclidean distance can be defined between any two data items. The clustering algorithm then functions based on an FCM-style objective function and produces clusteroids which are defined in the same feature space as the data [56]. These cluster prototypes may not be explicitly represented in the data space, but, nevertheless, they share the same mathematical model as the data [57] (the reader is referred to a review of Kernel FCM (KFCM) and Multiple-Kernel FCM (MKFCM) in [58] and several variants of KFCM in [59]). Another example for an intrinsically prototype-based clustering approach in which the prototypes are not explicitly "visible" is the Fuzzy PCA-guided Robust k-means (FPR k-means) clustering algorithm [60] in which a centroid-less formulation [61] is adopted which, nevertheless, defines homogeneity as proximity between the data items.

Relational clustering approaches constitute another class of algorithms which are intrinsically based on the distances between the data items [24]. The reader is referred to Relational FCM (RFCM) [62] and its non-Euclidean extension Nerf C-means [63] for examples. The goal of this class of algorithms is to separate the data into *self-similar* bunches. Another algorithm in which the presence of prototypes may be less evident is Multiple Prototype Fuzzy Clustering Model (FCMP) [64], in which the data items are described as a linear combination of a set of prototypes, which are, nevertheless, members of the same $\mathbb{R}^k$ as the data is. Fuzzy clustering by Local Approximation of Memberships (FLAME) [65] and Hierarchical Agglomerative Clustering (HAC) [66, 14.3.12 Hierarchical clustering] are other clustering algorithms which inherently guide the process of clustering based on the distances between the data items. The same is applicable to Visual Assessment of cluster Tendency (VAT) [67], and its variants Automated VAT (aVAT) [68] and Improved VAT (iVAT) [69], which all function based on the distances between the data items. Additionally, some researchers utilize $\ell_p$-norms, for $p \neq 2$ [70, 71, 72], or other distance functions which operate on pairs of data items [73].

We argue that a successful departure from the assumption of prototypical clustering is achieved when clusters and data items have different mathematical models. For example, the Gustafson-Kessel algorithm [74] models a cluster as a pair of a point and a covariance matrix and utilizes the Mahalanobis distance between data items and clusters (also see the Gath-Geva algorithm [75]). Fuzzy shell clustering algorithms [17] utilize more generic geometrical structures. For example, the FCV [54] algorithm can detect lines, planes, and other hyper-planar forms (also see [76, 77, 78, 79, 80, 81, 82]), Fuzzy C Ellipsoidal Shells (FCES) [83] searches for ellipses, ellipsoids, and hyperellipsoids, and Fuzzy C Quadric Shells (FCQS) [17] and its variants seek quadric and hyperquadric clusters.

## 2.3   Robustification

Dave et al. [84] argue that the concept of the membership function in FCM and the concept of the weight function in robust statistics are related. Based on this perspective, it is argued that the classical FCM in fact provides an indirect means for attempting robustness. Nevertheless, it is known that FCM and other least square methods are highly sensitive to noise [34]. Hence, there has been ongoing research on the possible modifications of the FCM in order to provide a (more) robust clustering algorithm [85, 86]. Dave et al. [84] provide an extensive list of relevant works and outline the intrinsic similarities within a unified view (also see [87]).

The first attempt to robustifying the FCM, based on one account [84], is the Ohashi Algorithm [88]. That work adds a noise class to FCM and generates a robustified objective function through utilizing the regularization term,

$$\tilde{\Delta} = \sum_{n=1}^{N}(1 - \sum_{c=1}^{C} f_{nc})^m.$$ (2.3)

That transformation was suggested independently by Dave [87, 89] when he developed the Noise Clustering (NC) algorithm as well. The core idea in NC is that there exists one additional imaginary prototype which is at a fixed distance from all the data items and represents noise.

Krishnapuram et al. [90] extended the idea behind NC and developed the Possibilistic C-means (PCM) algorithm by rewriting the objective function as (also see [91]),

$$\Delta = \sum_{c=1}^{C}\sum_{n=1}^{N} t_{nc}^m \|x_n - \psi_c\|^2 + \sum_{c=1}^{C} \eta_c \sum_{n=1}^{N}(1 - t_{nc})^m.$$ (2.4)

Here, $t_{nc}$ denotes the degree of representativeness or *typicality* of $x_n$ to $\psi_c$ (also addressed as *possibilistic degree* in contrast to the *probabilistic* model utilized in FCM). As expected from the modification in the definition of $t_{nc}$ compared to $f_{nc}$, PCM removes the sum of one constraint, shown in (2.2), and in effect extends the idea of one noise class in NC into $C$ noise classes. In other words, PCM could be considered as the parallel execution of $C$ independent NC algorithms, in which each seeks a cluster.

We argue that the interlocking mechanism present in FCM, i.e. (2.2), is valuable in that, not only clusters seek homogenous sets, but that they are also forced into more optimal "positions" through forces applied by competing clusters. In other words, borrowing the language used in [51], in FCM, clusters "seize" data items and it is disadvantageous for multiple clusters to claim high membership to the same data item. There is no phenomenon, however, in NC and PCM which corresponds to this internal factor. Additionally, it is likely that PCM clusters coincide and/or leave out portions of the data unclustered [92]. In fact, it is argued that the fact that at least some of the clusters generated through PCM are nonidentical is because PCM gets trapped into local minimum [93]. PCM is also known to be more sensitive to initialization than other algorithms in its class [51].

It has been argued that both concepts of possibilistic degrees and membership values have positive contributions to the purpose of clustering [94]. Hence, Pal et al. [95] combined FCM and PCM and rewrote the optimization function of Fuzzy Possiblistic C-Means (FPCM) as minimizing,

$$\Delta = \sum_{c=1}^{C} \sum_{n=1}^{N} (f_{nc}^m + t_{nc}^\eta) \left\| x_n - \psi_c \right\|^2, \tag{2.5}$$

subject to (2.2) and $\sum_{n=1}^{N} t_{nc} = 1, \forall c$. That approach was later shown to suffer from different scales of the $f_{nc}$ and $t_{nc}$ values, especially when $N \gg C$, and, therefore, additional linear coefficients and a PCM-style term were introduced to the objective function [96] (also see [97] for another variant). It has been argued that the resulting objective function employs four correlated parameters and that the optimal choice for them for a particular problem instance may not be trivial [51]. Additionally, in the new combined form, $f_{nc}$ cannot necessarily be interpreted as a membership value [51].

Weight modeling is an alternative robustification technique and is exemplified in the algorithm developed by Keller [98], where the objective function is rewritten as,

$$\Delta = \sum_{c=1}^{C} \sum_{n=1}^{N} f_{nc}^m u_c \frac{1}{\omega_n^q} \left\| x_n - \psi_c \right\|^2, \tag{2.6}$$

subject to $\sum_{n=1}^{N} \omega_n = \omega$. Here, the value of $\omega_n$ is updated during the process as well.

On a different tangent, Frigui et al. [16] included a robust loss function in the objective function of FCM and developed Robust C-Prototypes. They further extended RCP and developed an unsupervised version of RCP, nicknamed URCP [16]. Wu et al. [57] used $u_c(x) = 1 - e^{-\beta x^2}$ and developed Alternative HCM (AHCM) and Alternative FCM (AFCM) algorithms (also see [99]).

## 2.4   Number of Clusters

The classical FCM and PCM, and many of their variants, are based on the assumption that the number of clusters is known (an extensive review of this topic is given in [9, Chapter 4]). While PCM-style formulations may appear to relax this requirement, the corresponding modification is carried out at the cost of yielding an ill-posed optimization problem [51]. In fact, repeating the clustering for different numbers of clusters [75, 100] and *Progressive Clustering* are two of the alternative approaches to address the challenge of not requiring *a priori* knowledge about the number of clusters present in a particular data.

Among the many variants of Progressive Clustering are methods which start with a significantly large number of clusters and freeze "good" clusters [100, 101, 81, 82], approaches which combine compatible clusters [102, 100, 101, 81, 82, 16], and the technique of searching for one "good" cluster at a time until no more is found [38]. These approaches utilize one or more CVIs in order to assess the appropriateness of the clusters produced after each execution of the algorithm. For a review of CVIs in the context of relational clustering refer to [24] (also see [23]). Use of regularization terms in order to push the clustering results towards the "appropriate" number of clusters is another approach taken in the literature [103]. These regularization terms, however, generally involve additional parameters which are to be set carefully, and potentially per problem instance [95].

Dave et al. conclude in their 1997 paper that the solution to the general problem of robust clustering when the number of clusters is unknown is "elusive" and that the techniques available in the literature each have their limitations [84]. In this paper, we acknowledge that the problem of determining the appropriate number of clusters is hard to solve and even hard to formalize. Additionally, we argue that this challenge is equally applicable to many clustering problems independent of the particular clustering models utilized in the algorithms. Therefore, we designate this challenge as being outside the scope of this contribution and assume that either the appropriate number of clusters is known or that an exogenous means of cluster pruning is available which can be utilized within the context of the algorithm developed in this paper.

## 2.5 Weber Problem

In 1929, Weber published a pioneering work on finding an optimal solution to a problem which is now commonly known by his name [104]. In its modern form [105], the *Weber Problem* can be written as minimizing,

$$\Delta(\psi) = \sum_{n=1}^{N} \omega_n \phi(x_n, \psi). \tag{2.7}$$

Here, $\omega_n > 0$ are the weights and $\phi(\cdot)$ is a known function. When $\phi(\cdot)$ denotes the Euclidean distance, one of the most popular solution strategies to the Weber Problem is proposed by Weiszfeld [106] (refer to [107] for an accelerated version).

A generic approach to solving (2.7) is to confirm that $\Delta(\psi)$ is convex and then to pursue with the derivative of $\Delta(\cdot)$ relative to $\psi$. This process generally results in the presence of $\phi'(x_n, \psi)$ in the update equation of an iterative process, for which guaranteeing convergence is often cumbersome [108]. For example, the analysis given in [109] does not contain a proof of convergence and suggests to monitor the value of $\Delta(\psi)$ during the iterations and to restrict the rate of change in $\psi$. While this operation is costly, its generalization to problems in which $\psi$ is governed by a complex mathematical model, as opposed to $\psi \in \mathbb{R}^k$, is non-trivial. Other works in the field consider $\ell_p$ norms and discuss local and global convergence [110, 111] as well as recommend acceleration techniques [112, 113] (also see Iteratively Reweighted Least Squares (IRLS) [114]). The reader is referred to [115, Chapter 4.5] and the references therein for further review of the topic and different options for numerical calculation.

Other approaches to solving the Weber Problem in more general settings, utilize regularization and other similar numerical techniques in order to suppress the amount of change in $\psi$ in consecutive iterations. As such, the progression of the algorithm is maintained at a balance between a gradient descent path with smaller sizes, which is slow but guaranteed to converge, and a Newton's method, which is faster but may diverge. The reader is referred to Tikhonov-Arsenin [116], Levenberg [117], and Marquardt [118] algorithms for mathematical details (also see [119]). A very recent review of the different incarnations of the Weber Problem and an outline of Weiszfeld's work can be found in [120].

From a theoretical perspective, the Weber Problem is a special case of the problem addressed in this paper when $C = 1$. Nevertheless, as will be discussed in Section 3.1, the two functions $\Delta_{\mathbf{X}}(\psi)$ and $\Delta_{\mathbf{\Psi}}(x)$, which are defined and used in this work, have structural similarities to generalized

Weber Problems. Therefore, some of the tools developed for solving generalized Weber Problems are applicable to the method developed in this paper.

# Chapter 3

# Developed Method

## 3.1 Model Preliminaries

We assume that a clustering problem class is given, within the context of which a model for data items is known and denote a data item as $x$. We also assume that the provided problem class defines a cluster model, which complies with the notion of homogeneity relevant to the problem class at hand, and denote a cluster as $\psi$. We note that the model utilized in this work is based on previous works in the field [121].

In this work, we utilize a weighted set of data items, defined as,

$$\mathbf{X} = \Big\{ (\omega_n; x_n) \Big\}, n = 1, \cdots, N, \omega_n > 0, \tag{3.1}$$

and we define the *weight* of $\mathbf{X}$ as $\Omega(\mathbf{X}) = \sum_{n=1}^{N} \omega_n$. When known in the context, we abbreviate $\Omega(\mathbf{X})$ as $\Omega$. Thus, when estimating expected values, we treat $\mathbf{X}$ as a set of realizations of the random variable $x$ and write $p\{x_n\} = \frac{\omega_n}{\Omega}$.

We assume that the real-valued positive *distance function* $\phi(x, \psi)$ is defined. Through this abstraction, we allow for any applicable distance function and therefore decidedly avoid the dependence of the underlying algorithm on Euclidean or any other particular notations of distance. As examples, when the data items belong to $\mathbb{R}^k$, the Euclidean Distance, any $\ell_p$ norm, and the Mahalanobis Distance are special cases of the notion of data item to cluster distance defined here. The corresponding cluster models in these cases denote $\psi \in \mathbb{R}^k$, $\psi \in \mathbb{R}^k$, and $\psi$ identifying a pair of a member of $\mathbb{R}^k$ and a $k \times k$ covariance matrix, respectively.

We assume that $\phi(x, \psi)$ is differentiable in terms of $\psi$ and that for any non-empty weighted set

13

**X** the following function of $\psi$,

$$\Delta_{\mathbf{X}}(\psi) = E\{\phi(x, \psi)\} = \frac{1}{\Omega} \sum_{n=1}^{N} \omega_n \phi(x_n, \psi), \tag{3.2}$$

has one and only one minimizer which is also the only solution to the following equation,

$$\sum_{n=1}^{N} \omega_n \frac{\partial}{\partial \psi} \phi(x_n, \psi) = 0, \tag{3.3}$$

In this paper, we assume that a function $\Xi_\psi(\cdot)$ is given, which, for the input weighted set **X**, produces the optimal $\psi$ which minimizes (3.2) and is the solution to (3.3). We address $\Xi_\psi(\cdot)$ as the *cluster fitting function*. Examples for $\Xi_\psi(\cdot)$ include the mean and the median when $x$ and $\psi$ are real values and $\phi(x, \psi) = (x - \psi)^2$ and $\phi(x, \psi) = |x - \psi|$, respectively. Note the similarity between (3.2) and (2.7).

We also assume that $\phi(x, \psi)$ is differentiable in terms of $x$ and that for the input set $\boldsymbol{\Psi} = \{(\omega_c; \psi_c)\}, c = 1, \cdots, C, \omega_c > 0$, the following function of $x$,

$$\Delta_{\boldsymbol{\Psi}}(x) = \frac{1}{\Omega} \sum_{c=1}^{C} \omega_c \phi(x, \psi_c), \tag{3.4}$$

has one and only one minimizer which is also the only solution to the following equation,

$$\sum_{c=1}^{C} \omega_c \frac{\partial}{\partial x} \phi(x, \psi_c) = 0, \tag{3.5}$$

Here, we assume that a function $\Xi_x(\cdot)$ is given, which, for the input weighted set $\boldsymbol{\Psi}$, produces the optimal $x$ which minimizes (3.4) and is the solution to (3.5). We address $\Xi_x(\cdot)$ as the *data responsiveness function* and accept that in certain cases, $\Xi_x(\cdot)$ may depend on $x$ as well. Again, note the similarity between (3.4) and (2.7).

In fact, both $\Xi_\psi(\cdot)$ and $\Xi_x(\cdot)$ are solutions to the M-estimators given in (3.2) and (3.4), respectively. We note that when a closed-form representation for either $\Xi_\psi(\cdot)$ or $\Xi_x(\cdot)$ is not available, conversion to W-estimators can produce procedural solutions to either (3.3) or (3.5), respectively [122]. Additionally, many of the techniques developed in the context of the Weber Problem may be applicable to these two functions as well.

We assume that a function $\Xi_o(\cdot)$, which may depend on **X**, is given, that produces an appropriate number of initial clusters. We address this entity as the *cluster initialization function* and denote the number of clusters produced by it as $C$. In this work, as discussed in Section 2.4, we assume that the management of the value of $C$ is carried out by an external process which may also intervene between the successive iterations of the algorithm developed in this paper.

We assume that a robust loss function, $u(\cdot) : [0, \infty] \rightarrow [0, 1]$, is given which satisfies $\lim_{\tau \rightarrow \infty} u(\tau) = 1$. Additionally, we assume that $u(\cdot)$ is an increasing differentiable function which satisfies $u(0) \simeq 0$ and $u(1) \simeq \frac{1}{2}$.

In this work, we utilize the rational robust loss function given below,

$$u(x) = \frac{x + \varepsilon}{1 + x}, 0 < \varepsilon \ll 1. \tag{3.6}$$

Here, in order to avoid numerical instability at $x = 0$, $\varepsilon = 2 \times 10^{-16}$ is utilized. Derivation shows that $u'(x) = (1 - \varepsilon)(1 + x)^{-2}$ and $u''(x) = -2(1 - \varepsilon)(1 + x)^{-3}$, i.e. $u(x)$ is an increasing concave function.

Using (3.6), we model the loss of $x_n$, when it belongs to $\psi_c$, as,

$$u_{nc} = u\left(\frac{1}{\lambda}\phi_{nc}\right), \phi_{nc} = \phi(x_n, \psi_c). \tag{3.7}$$

We address $\lambda$ as the *scale* parameter (note the similarity with the cluster-specific weights in PCM [90]). In fact, $\lambda$ has a similar role to that of scale in robust statistics [123] (also called the *resolution* parameter [53]) and the idea of distance to noise prototype in the NC algorithm [87, 89]. Scale can also be considered as the controller of the boundary between inliers and outliers [84]. From a geometrical perspective, $\lambda$ controls the radius of spherical clusters and the thickness of planar and shell clusters [51]. One may investigate the possibility of generalizing this single identity into cluster-specific scale factors, i.e. $\lambda_c$ values, in line with the $\eta_c$ variables in PCM [90].

We also model the loss of a data item which is considered to be an outlier as $U > 0$. In this paper, we propose two deterministic procedures which produce the values of $\lambda$ and $U$ in a particular problem class independent of $\mathbf{X}$.

## 3.2    Assessment of Loss

The problem addressed in this paper is to find $C$ locally optimal clusters which comply with the notion of homogeneity identified through $\phi(\cdot)$. We produce the set of fuzzy membership values $f_{nc} \in [0, 1]$, where $f_{nc}$ denotes the membership of $x_n$ to $\psi_c$ (exact definition to be given later) and satisfies (2.2). The algorithm developed in this paper also produces the *possibility* identifiers $p_n \in [0, 1]$, which denote the algorithm's estimate of the possibility that $x_n$ is an inlier. To put in perspective, as will be discussed later, $p_n f_{nc}$ in this work is comparable to $f_{nc}$ in FCM. The $p_n$ identifiers in this work have a role distantly comparable to that of the $t_{nc}$ in PCM.

16

Here, we first derive a loss model for the described clustering problem. To do this, we employ a Bayesian inference framework and model the aggregate loss of an arbitrary solution to the clustering problem. In the following derivations, conditional probabilities are dropped when they are trivial in the context.

$$E\left\{\text{Loss}|\mathbf{X}\right\} = \sum_{n=1}^{N} E\left\{\text{Loss}\middle|x_n\right\} p(x_n) = \tag{3.8}$$

$$\frac{1}{\Omega}\sum_{n=1}^{N} \omega_n E\left\{\text{Loss}\middle|x_n\right\} =$$

$$\frac{1}{\Omega}\sum_{n=1}^{N} \omega_n \left[ E\left\{\text{Loss}\middle|x_n \in \tilde{\mathbf{X}}\right\} p\left\{x_n \in \tilde{\mathbf{X}}\right\} + E\left\{\text{Loss}\middle|x_n \notin \tilde{\mathbf{X}}\right\} p\left\{x_n \notin \tilde{\mathbf{X}}\right\} \right].$$

Here, $\tilde{\mathbf{X}}$ is the subset of $\mathbf{X}$ which only includes the inliers. We model the possibility that $x_n$ is an inlier as $p_n$. Therefore,

$$E\left\{\text{Loss}|\mathbf{X}\right\} = \frac{1}{\Omega}\sum_{n=1}^{N} \omega_n \left[ p\left\{x_n \in \tilde{\mathbf{X}}\right\} \right. \tag{3.9}$$

$$\sum_{c=1}^{C} E\left\{\text{Loss}\middle|x_n \in \tilde{\mathbf{X}}_c\right\} p\left\{x_n \in \tilde{\mathbf{X}}_c\middle|x_n \in \tilde{\mathbf{X}}\right\} +$$

$$\left. E\left\{\text{Loss}\middle|x_n \notin \tilde{\mathbf{X}}\right\} p\left\{x_n \notin \tilde{\mathbf{X}}\right\} \right] =$$

$$\frac{1}{\Omega}\sum_{n=1}^{N} \omega_n \left[ p_n \sum_{c=1}^{C} u_{nc} f_{nc} + C\frac{1}{C}U(1-p_n) \right].$$

In this model, $\tilde{\mathbf{X}}_c$ is the subset of $\tilde{\mathbf{X}}$ which contains the data items which belong to $\psi_c$. Moreover, $f_{nc}$ denotes the probability that $x_n$ belongs to $\psi_c$, given that it is an inlier. Here, for reasons which will become clear later, we have chosen to rewrite the constant 1 as $C\frac{1}{C}$.

## 3.3  Fuzzification

Close assessment of (3.9) shows that this cost function contains an HCM-style hard formulation. It is known, however, that the transformation of an HCM-style clustering objective function to an alternative fuzzy/probabilistic formulation has important benefits, as outlined in Section 2.1. Hence, we rewrite (3.9) and derive the cost function for the clustering problem developed in this paper as,

$$\Delta = \sum_{c=1}^{C} \sum_{n=1}^{N} \omega_n f_{nc}^m p_n^m u_{nc} + U C^{1-m} \sum_{n=1}^{N} \omega_n \left(1 - p_n\right)^m. \tag{3.10}$$

This objective function is to be minimized subject to (2.2).

A note on the concept of membership in the present framework is necessary. As stated before, in this work, $f_{nc}$ denotes the membership of $x_n$ to $\psi_c$ conditional to $x_n$ being an inlier. In other words, the membership of $x_n$ to $\psi_c$, as commonly referenced in the literature, is in fact equal to $p_n f_{nc}$.

We emphasize that the second term in (3.10) in fact acts as a regularization component and that it bears resemblance to the regularization term in PCM (see (2.4)). It is important to emphasize, however, that the second term in (3.10) is in fact derived organically through the derivation of the loss value. In comparison, a significant majority of the well-known methods in the literature, including PCM, utilize regularization terms which are heuristically and based on the intuition of the researchers assumed to yield the desired effect (for example see [124]). Those works then utilize experimental evidence in order to exhibit that the desired effects are in fact present and valid. In contrast, in this work, we start from a Bayesian model for the loss function and derive an objective function which contains a component that may be understood as a regularization term. We find this distinction important from an epistemological perspective. Additionally, as will be discussed later, in this work, the value of $U$ is set through a deterministic procedure. This is in clear contrast with previous works which require the proper setting of a regularization coefficient by the user based on intuition, expertise, or multiple repetitions of the experiment.

## 3.4   Problem Development

Implementing (2.2) in (3.10) through Lagrange Multipliers and equating the partial derivative of the cost function in terms of $f_{nc}$ to zero, yields,

$$f_{nc} = \frac{u_{nc}^{-\frac{1}{m-1}}}{\sum_{c'=1}^{C} u_{nc'}^{-\frac{1}{m-1}}}. \tag{3.11}$$

18

Subsequently, optimizing the cost function in terms of $p_n$ gives,

$$p_n = \frac{\sum_{c=1}^{C} u_{nc}^{-\frac{1}{m-1}}}{\sum_{c=1}^{C} u_{nc}^{-\frac{1}{m-1}} + CU^{-\frac{1}{m-1}}}.$$

(3.12)

Now, we insert (3.11) and (3.12) in (3.10) and derive,

$$\Delta = \sum_{n=1}^{N} \omega_n \left[ \sum_{c=1}^{C} u_{nc}^{-\frac{1}{m-1}} + CU^{-\frac{1}{m-1}} \right]^{-(m-1)}.$$

(3.13)

Note that this objective function is independent of any terms except for the cluster representation $\psi_1, \cdots, \psi_C$ and is to be minimized unconstrained to any constraints.

As noted in Section 2.1, in this work, we use $m = 2$. Hence, (3.13) can be rewritten in the simplified form,

$$\Delta = \sum_{n=1}^{N} \omega_n \frac{1}{\sum_{c=1}^{C} \frac{1}{u_{nc}} + C\frac{1}{U}}.$$

(3.14)

## 3.5   Solution Strategy

We equate the partial derivative of $\Delta$ with respect to $\psi_c$ with zero and produce $\sum_{n=1}^{N} \omega_{nc}^x \frac{\partial}{\partial \psi_c} \phi(x_n, \psi_c) = 0$. Here,

$$\omega_{nc}^x = \omega_n \left[ \sum_{c'=1}^{C} u_{nc'}^{-\frac{1}{m-1}} + CU^{-\frac{1}{m-1}} \right]^{-m} u_{nc}^{-\frac{m}{m-1}} u_{nc}',$$

(3.15)

where,

$$u_{nc}' = u' \left( \frac{1}{\lambda} \phi(x_n, \psi_c) \right).$$

(3.16)

Now, utilizing the definition of $\Xi_\psi(\cdot)$ we infer that,

$$\psi_c^\star = \Xi_\psi \left( \left\{ (\omega_{nc}^x; x_n) \right\} \right).$$

(3.17)

This equation leads to an iterative procedure which updates the $\psi$ representations. Convergence considerations for this process are addressed in Section 3.9. Note that, when $m = 2$ and for (3.6), (3.15) reduces to,

$$\omega_{nc}^x = \omega_n \left( \frac{1}{\lambda} \phi_{nc} + \varepsilon \right)^{-2} \left( \sum_{c'=1}^{C} \frac{1}{u_{nc'}} + C\frac{1}{U} \right)^{-2}.$$

(3.18)

In this work, we propose to postpone the calculation of the $f_{nc}$ and $p_n$ values until convergence is achieved. In fact, as will be demonstrated in Section 3.6, the calculation of the $f_{nc}$ and $p_n$ values may never be necessary in many settings. Nevertheless, (3.11) and (3.12) can be utilized accordingly, if the $f_{nc}$ and $p_n$ are to be produced. We emphasize that because this approach eliminates the need for updating the $f_{nc}$ and $p_n$ values during the process, it reduces the memory load of the algorithm from $(2C + 1)N$, for $u_{nc}$, $f_{nc}$ and $p_n$, to $CN$, for $u_{nc}$.

## 3.6 Classification & Outlier Removal

In order to perform classification, we propose to utilize a Maximum Likelihood inference framework and to compare the probability values corresponding to $x_n \in \tilde{\mathbf{X}}_c$, for $c = 1, \cdots, C$, as well as the probability that $x_n \in \mathbf{X} - \tilde{\mathbf{X}}$. Hence, a data item belongs to one of the $\tilde{\mathbf{X}}_c$, and therefore is an inlier, when $p_n \max_c f_{nc}$ is bigger than $1 - p_n$. Utilizing (3.11) and (3.12) one can demonstrate that this relationship can be written as,

$$\min_c u_{nc} < C^{-(m-1)}U. \tag{3.19}$$

Therefore, one does not need to calculate the $f_{nc}$ and $p_n$ values in order to determine which cluster a certain data item belongs to and whether or not it is an outlier. In summary, if the problem is to be solved in *inclusive* mode, $x_n$ is assigned to $\psi_{c_n}$, where,

$$c_n = \arg\min_c u_{nc}. \tag{3.20}$$

Here, $c_n$ is the cluster to which the data item $x_n$ is assigned. In *non-inclusive* mode, however, $c_n$ is not defined if $x_n$ is not an inlier, i.e. if it does not satisfy (3.19).

## 3.7 Data Responsiveness

Clustering algorithms have historically tried to satisfy two aims, where different embodiments approach either or both of these goals. On the cluster side, the purpose is to identify a number of clusters which best represent the data. On the data side, however, the algorithms are interested in the separation of the data into homogenous sets. As discussed in Section 3.6, in the algorithm developed in this paper, both of these goals are targeted and, more importantly, there is no need to calculate the $f_{nc}$ and $p_n$ values in order to satisfy these requirements. This is an important contribution of the developed algorithm from a practical perspective. Additionally, in the present work, we develop the process of allowing the data to actively respond to the clusters.

Utilizing (3.13), we propose to reassign the roles in the problem and to require the optimization of $x_n$ given known $\psi_c$. As such, the clustering problem is rephrased as follows: How can the data items be modified in order to be better members of the collection of homogenous sets that they are assigned to according to the given fuzzy membership values. Intuition may suggest that, for example, in the case of hyper-dimensional clusters, one should attempt to project each data item into the representative of the cluster that they are assigned to. However, as has been the approach of this paper, we avoid such model-specific heuristics and formalize the problem as minimizing (3.13) for $x_n$.

Equating the partial derivative of a modified version of $\Delta$ relative to $x_n$ to zero yields,

$$\sum_{c=1}^{C} \omega_{nc}^{\psi} \frac{\partial}{\partial x_n} \phi(x_n, \psi_c) = 0, \tag{3.21}$$

Here,

$$\omega_{nc}^{\psi} = u_{nc}^{-\frac{m}{m-1}} u_{nc}'. \tag{3.22}$$

Utilizing the definition of $\Xi_x(\cdot)$, we infer from (3.21) that,

$$x_n^{\star} = \Xi_x \left( \left\{ \left( \omega_{nc}^{\psi}; \psi_c \right) \right\} \right). \tag{3.23}$$

Note that this equation leads to an iterative procedure which updates the data items. Convergence considerations for this process are addressed in Section 3.9. We recommend that in non-inclusive settings, (3.23) is only applied on inliers.

Note that, when $m = 2$ and for (3.6), (3.22) reduces to $\omega_{nc}^{\psi} = \left( \frac{1}{\lambda} \phi_{nc} + \varepsilon \right)^{-2}$.

We emphasize that, in the present work, data responsiveness is only allowed to happen after the clustering process has converged. This is a choice made in order to postpone computational difficulties and convergence concerns. We recognize that one may investigate the possibilities enabled through data responsiveness midst process. Under such a regime, while clusters traverse towards a locally optimal configuration, the data items will also evolve, possibly facilitating algorithm convergence.

We also note that the process of data responsiveness has no impact on neither the cluster representations nor the relationship between the clusters and the data items. Additionally, data responsiveness is an optional component of the algorithm developed in this paper. Hence, when

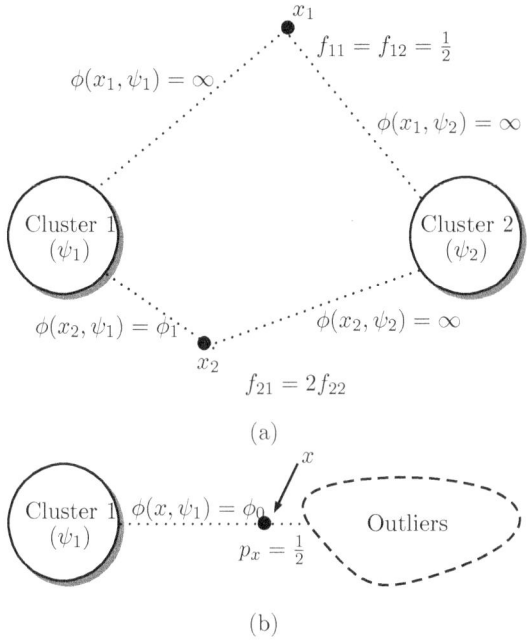

Figure 3.1: The process of determining the values of $\lambda$ and $U$. (a) $\lambda$. (b) $U$.

the clustering process is executed for the purposes of listing the clusters relevant to the data items or to perform data item classification, the process of data responsiveness is in effect irrelevant to the outputs. In other words, the fact that the present work allows for data responsiveness is an optional contribution over present works in the literature, and not an alteration of the goal of data clustering.

## 3.8  Determination of $U$ and $\lambda$

It is evident that $\lambda$ defines the scale for $\phi_{nc}$. This is exemplified in (3.7) and also everywhere else in this paper where $\phi_{nc}$ is divided by $\lambda$. We argue that, similarly, $U$ defines the scale for $u_{nc}$. Hence, we argue that the two identities $\phi_{nc}$ and $u_{nc}$ are brought into context through $\lambda$ and $U$, respectively. We use this perceptual definition in order to propose a procedure for determining the appropriate values for $\lambda$ and $U$ for a particular problem class.

We suggest an imaginary situation, as depicted in Figure 3.1-(a), in which two data items interact with two clusters ($N = 2$ and $C = 2$). The first data item, here $x_1$, is infinitely far from both clusters, in which case we expect $f_{11} = f_{12} = \frac{1}{2}$. This equilibrium is in fact verified by (3.11). The second data item, here $x_2$, however, is infinitely far from the second cluster and at a distance of $\phi_1$ from the first one. Here, we ask what value of $\phi_1$ will result in $f_{21} = 2f_{22}$. This question can be asked in a different setting as follows: For a data item which is infinitely far from two clusters, how close should it get to one cluster, while maintaining its distance to the other, in order to be favored by the former cluster two times than the latter?

Utilizing (3.11) one can show that $\lambda = \dfrac{1}{u^{-1}\left(2^{-(m-1)}\right)}\phi_1$. Considering the special case utilized in the experiments carried in this paper, i.e. $m = 2$, one can show that independent of the choice of $u\left(\cdot\right)$, we have $\lambda \simeq \phi_1$.

In order to estimate the proper value for $U$, we utilize another imaginary situation in which one data item interacts with one cluster, as depicted in Figure 3.1-(b). Here, we ask how far the data item should be from the cluster in order for it to be an inlier with a probability of half? This situation, in effect, defines the boundary of inliers and outliers when Maximum Likelihood is applied on $p_n$. We denote this distance as $\phi_0$ an derivation using (3.12) shows that $U = u\left(\frac{1}{\lambda}\phi_0\right)$. It is important to emphasize that the outcome of this process is independent of $C$. In other words, in a set of data items which contains $C$ clusters, a data item which is at the same distance of $\phi_0$ from all of them will be an inlier with the probability of half.

## 3.9 Convergence

The process developed in this paper utilizes two iterative procedures, i.e. (3.17) and (3.23). These two procedures are both techniques used in order to solve the generalized Weber Problem given in (3.13) from two different perspectives.

We emphasize that this contribution does not provide a proof that the application of either or both of (3.17) and (3.23) results in a decline of $\Delta$. Instead, we utilize the practical solution of monitoring the value of $\Delta$ after each step and reverting to the previous configuration if the execution of either (3.17) or (3.23) in fact inflates $\Delta$.

One may replace this primitive approach with one of the approaches discussed in Section 2.5. Alternatively, one may investigate special conditions on $u\left(\cdot\right)$ and $\phi\left(\cdot\right)$ which guarantee convergence for (3.17) and (3.23).

## 3.10 Implementation Notes

Algorithm 1 shows the outline of the implementation of the method developed in this paper. This algorithm is implemented as a class named *Grace* in MATLAB Version 8.1 (R2013a). This code takes use of Image Processing Toolbox for minor image-related primitives and the major operations are implemented as C/MEX dll's. The code is executed on a Personal Computer which runs Windows 7, 64bit, on an Intel Core i5-2400 CPU, 3.10GHz, with 8.00GB of RAM.

Each problem class is implemented as a child class for *Grace*. The child classes implement a constructor which creates the weighted set $\mathbf{X}$ based on the input image, data file, etc. The child classes also implement the four functions $\phi(\cdot)$, $\Xi_\circ(\cdot)$, $\Xi_\psi(\cdot)$, and $\Xi_x(\cdot)$. Hence, in order to define a new problem class, it suffices to define the aforementioned functions and to implement proper data-loading and output-visualization routines. In fact, the child classes are not responsible for any of the core operations of the developed algorithm. These operations are implemented in the parent class.

**Input:**

- Weighted Set of Data Items: $\mathbf{X}$

**Output:**

- Cluster Representations: $\psi_1, \cdots \psi_C$

- Outlier Decisions: $p_n$ [*]

- Classification Results: $f_{nc}$, $c_n$ [*]

- Updated Set of Data Items: $\mathbf{X}$ [*]

Call $\Xi_\circ(\mathbf{X})$ in order to produce $\psi_1, \cdots \psi_C$;

**while** *True* **do**

    **for** $n = 1$ *to* $N$, $c = 1$ *to* $C$ **do**

        Calculate $\phi_{nc}$ and $u_{nc}$ using (3.7);

        Calculate $u'_{nc}$ using (3.16);

    **end**

    Calculate $\Delta$ using (3.13);

    **for** $c = 1$ *to* $C$ **do**

        **for** $n = 1$ *to* $N$ **do** Calculate $\omega^x_{nc}$ using (3.15);

        Calculate $\psi^\star_c$ using (3.17);

        Assume that $\psi_c$ is updated to $\psi^\star_c$ and calculate the new value of $\Delta$, using (3.13), and address it as $\Delta^\star$;

        **if** $\Delta^\star < \Delta$ **then** Update $\psi_c$ to $\psi^\star_c$;

    **end**

    **if** *significant change in $\Delta$ is not registered in multiple iterations* **then** Break;

**end**

// Run if classification and outlier removal are required.

**for** $n = 1$ *to* $N$ **do**

    Perform outlier removal for $x_n$ using (3.19);

    Calculate $c_n$ using (3.20);

**end**

// Run if membership values are required.

**for** $n = 1$ *to* $N$ **do**

    Calculate $p_n$ using (3.12);

    **for** $c = 1$ *to* $C$ **do** Calculate $f_{nc}$ using (3.11);

**end**

// Run if data responsiveness is required.

**for** $n = 1$ *to* $N$ **do**

    **for** $c = 1$ *to* $C$ **do** Calculate $\omega^\psi_{nc}$ using (3.22);

    Update $x_n$ using (3.23);

**end**

**Algorithm 1:** Outline of the algorithm developed in this paper. Outputs denoted by * are optional.

# Chapter 4

# Experimental Results

This paper proposes both a novel clustering algorithm and a novel concept within the well-established realm of data clustering, i.e. data responsiveness. Additionally, here, we argue against a conventional validation paradigm in the data clustering literature. These three facets of the present work are reviewed in this section through verbal discussions as well as experimental evaluation. First, we provide some notes on validation of clustering algorithms.

It is a common practice in the literature that the proposal of a novel clustering algorithm is complemented with experimental results which exhibit the usefulness and superior performance of the new algorithm compared to the previous ones. In this line of argument, the clustering methodology is often given *as is* and verbal arguments are given for its appropriateness and value. Subsequently, experimental results are provided in order to produce after-the-fact justification for the intuition of the researchers.

We argue that this approach can be challenged from at least two perspectives. First, from an epistemological vantage point, the methodology utilized for constructing a clustering algorithm must be clearly stated and it does not suffice to provide objective functions and constrains which "make sense". Additionally, from the perspective of verification methodologies, we argue that for an algorithm which contains multiple parameters, one can always fine-tune the process in order for it to function "better" than any other algorithm for a certain subset of inputs. This is in addition to the fact that even a large collection of successful executions do not provide justification for the appropriateness of an algorithm for the general case. As discussed before, when the aforementioned algorithms involve multiple parameters which are to be set empirically or based on heuristics, the situation is more troubling.

We propose that the value of a clustering algorithm can be assessed using the combination of two methods. One, the mathematical model on which the clustering problem is based must be constructed based on assumptions and a methodology which are explicitly and clearly stated. Hence, any model variable, parameter, constraint, or objective function must be clearly defined and ambiguous verbal descriptions and metaphors must be avoided. Two, the algorithm must be applied on a number of problem classes and instances and the results must be acceptable. While the first method provides theoretical justification, the second one delivers practical justification. This duo can be further complemented when the algorithm and its competitors are applied on a large set of standard data sets, the results of which must be examined and compared.

In this paper, we provided theoretical justification for the developed method in Chapter 3. Hence, to follow the argument given in the above, we outline the execution of the developed algorithm for problem instances in three different problems classes which concern 1D, 2D, and 3D data items. We review the experimental results collected through this effort in Section 4.2. Additionally, as discussed before, this contribution proposes the novel concept of data responsiveness in the realm of data clustering. In effect, to the best of our understanding, previous works have always considered data items as passive entities which are subjected to clustering, without responding to the results. Here, we describe the clustering problem as a mutual relationship between clusters and data items. Hence, while clusters in this work traverse towards being better fits for the available data, as it is the case with any clustering algorithm, we also allow the data items to evolve into better fits for the sets of clusters. In Section 4.3 we demonstrate the practical implications of this concept. Finally, the present work does not provide proof of convergence for two of the steps in the process. We acknowledge that only a rigorous mathematical proof will relax concerns regarding the convergence of the algorithm. Nevertheless, we present the results of a study on the convergence of the proposed method in Section 4.4. These results indicate that the value of the cost function may indeed inflate during the process, but that such an incident is both unlikely and insignificant in magnitude.

## 4.1  Models

The algorithm proposed in this work is class independent. Hence, in effect, the data items and the clusters may follow any arbitrary model which mutually comply. Here, in order to assess the performance of the proposed method, we review the application of the developed algorithm on three problem classes which concern 1D, 2D, and 3D data items, in Sections 4.1.1, 4.1.2, and 4.1.3, respectively.

Note that, as discussed in Section 3.10, the deployment of the proposed algorithm for these problem classes in fact only involves the implementation of a number of data manipulation functions for the particular data item and cluster models. In other words, the essence of the algorithm is shared between these deployments. Additionally, from a parameter-tuning perspective, the execution of the proposed algorithm for these problem classes involves no modification of any parameters, except for $\lambda$ and $U$, for which construction mechanisms are given in Section 3.8.

## 4.1.1 Grayscale Image Multi-Level Thresholding

The problem of grayscale image multi-level thresholding defines the data items as grayscale values and models a cluster as an interval on the grayscale axis centered at the scalar $\psi_c$ [38].

In order to produce the data, we calculate the histogram of the input image (32 bins). Here, each bin represents one data item and the number of pixels in the bin, normalized over the area of the image, produces the corresponding weight. Distance between a data item and a cluster is defined as the squared difference and the initial clusters are defined as uniformly-distributed points in the working range. Both $\Xi_\psi(\cdot)$ and $\Xi_x(\cdot)$ in this case calculate the weighted mean of the input scalars. We solve this problem as inclusive and utilize a difference of 32 graylevels as both $\phi_0$ and $\phi_1$. This value is selected as $\frac{1}{8}$ of the dynamic range of the signal.

## 4.1.2 2D Euclidean Clustering

The problem of 2D Euclidean Clustering is the classical case of finding dense clusters in weighted 2D data. Here, dense is defined in terms of inter-cluster proximity of points.

In this problem class, $\phi(\cdot)$ is the Euclidean distance and both $\Xi_\psi(\cdot)$ and $\Xi_x(\cdot)$ correspond to the weighted center of mass operator. The input set of data items in this experiment is generated based on the superposition of multiple Gaussian distributions and this problem class is defined as non-inclusive. While data items are inside the $[-12, 12] \times [-12, 12]$ square, we utilize 10 and 5 units of change as $\phi_0$ and $\phi_1$, respectively. These values are based on a $\chi_2^2$ model for the output of $\phi(\cdot)$ within the context of the data generation process.

### 4.1.3 Plane Finding in Range Data

This problem class is concerned with finding planar sections in range data. The input data in this problem class contains 3D points captured by a *Kinect 2* sensor. The depth-maps used in this experiment are captured at the resolution of $424 \times 512$ pixels. Here, intrinsic parameters of the camera are acquired through the Kinect SDK and each data item in this problem class has the weight of one.

Clusters in this problem class are defined as planes which have a thickness and the process of fitting a plane to a weighted set of 3D points is carried out through a weighted variant of Singular Value Decomposition (SVD). The data responsiveness function in this case respects the projective nature of the data and, therefore, attempts to reduce $\Delta$ while moving points along the corresponding epipolar lines. Hence, in this case, $\Xi_x(\cdot)$ depends on both $\Psi$ as well as $x$. Direct derivation shows that in this case,

$$\Xi_x\left(\Psi, \vec{x}\right) \equiv \frac{\displaystyle\sum_{c=1}^{C} \omega_c \vec{\psi}_c^T \vec{x}}{\displaystyle\sum_{c=1}^{C} \frac{\omega_c}{\|\vec{\psi}_c\|^2} \left(\vec{\psi}_c^T \vec{x}\right)^2} \vec{x}. \tag{4.1}$$

This problem is defined as non-inclusive, because we expect outliers and other data items which do not belong to any of the planes to exist in the data and wish to identify and exclude them.

In [125], the author utilizes the distance threshold of 20 $mm$ in order to determine whether a point is an inlier when running RANdom SAmple Consensus (RANSAC) [126]. The data acquisition process in that work utilizes an ASUS Xtion Pro Live device which is equivalent to the RGB-D sensor present in Microsoft Kinect. We multiply this threshold by ten, in order to address the presence of outliers and multiple clusters, and utilize $\phi_0 = \phi_1 = 200$ $mm$.

## 4.2 Comparative Results

In this section, we compare the performance of the proposed method with FCM. We have selected FCM for this comparison for two reasons. In effect, FCM is both self-sufficient and functional. In other words, FCM, in contrast to the majority of its descendants, does not depend on any parameter or regularization coefficient and additionally it produces acceptable results under many circumstances.

The presence of externally adjusted parameters is an important challenge, both practically

Figure 4.1: Standard image *House*.

and theoretically, in an evaluation process. As discussed in the beginning of this section, it is imaginable that a method which relies on several configuration parameters can be fine-tuned to function properly for many given problem classes. Nevertheless, this process, while tedious and resource expensive, is not a proof of validity. FCM, in this context, is a self-sufficient algorithm. Here, we review clustering results corresponding to the models outlined in Section 4.1.

### 4.2.1  Grayscale Image Multi-Level Thresholding

Figure 4.2 compares the results of applying FCM as well as the proposed method on the standard image *House*, shown in Figure 4.1. Here, the case of $C = 2$ is exhibited. In Figure 4.2, the top images visualize the results of the respective clustering algorithms for the data item contained in the image. In effect, the output of FCM is a $C$-level image, in which $x_n$ is substituted with $\psi_{c_n}$. In contrast, the output of the proposed method is a continuous field in which $x_n$ is replaced with $x_n^\star$. This is an explicit result of the concept of data responsiveness proposed in this paper. We present experimental results for data responsiveness in Section 4.3.

The second row in Figure 4.2 demonstrates data items and clusters in the data domain. Here, the thick dashed lines indicate the histogram of the input image, i.e. values of $\omega_n$, and the thick colored lines denote the membership of the data items to the clusters, where each color represents one cluster. The thin dashed lines in these charts denote the values of $p_n$. As expected from the FCM model, $p_n$ is always one, i.e. every data item is an inlier with a probability of one.

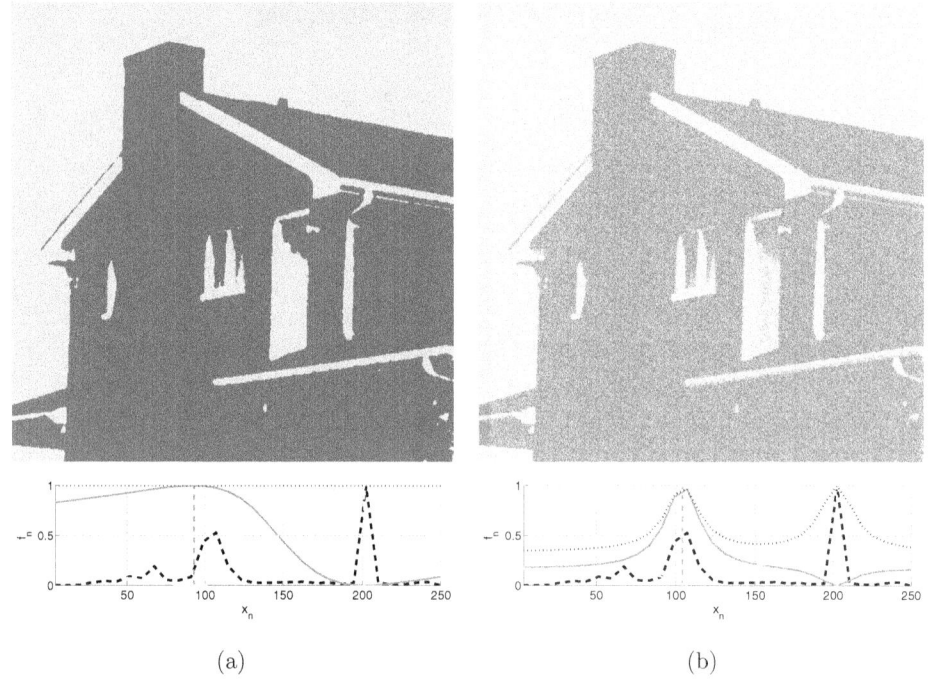

(a)  (b)

Figure 4.2: Results of applying FCM and the developed method for grayscale image multi-level thresholding on the Standard image *House*, shown in Figure 4.1, for $C = 2$. (a) FCM. (b) Proposed Method.

Comparison of both the top and the bottom rows in Figure 4.2 shows that the two algorithms converge to comparable results. Closer observation, however, yields a number of important differences. We particularly observe these differences between the two algorithms on the bottom row charts in Figure 4.2.

Examining patterns of membership to the clusters in the outputs of the two algorithms, we observe that FCM clusters are wider. In other words, membership graphs for FCM fall slowly. This is a direct result of (2.2), which mandates that at the middle point between two clusters, $f_{nc} = \frac{1}{2}$, independent of the distance between the two clusters. In other words, in FCM, the shapes of the membership functions are controlled exclusively by the mutual distances between the clusters. In the proposed method, however, membership to a cluster depends on $p_n$ as well and the scale parameter, $\lambda$, forces the membership charts to drop, even if no other competing cluster is present in the proximity. We observe this phenomenon in the area between the two clusters in the bottom row graphs in Figure 4.2 as well as the to the left and to the right of both graphs. In these areas, where no contestant clusters are present, FCM clusters claim high membership to distant data items. In contrast, the clusters generated by the proposed method drop the membership assigned to these data items. In fact, it is informative to review the geometry of the membership graphs for $x_n \in [0, 50]$ for both algorithms. These data items are in fact far from both clusters. Nevertheless, in FCM, the cluster closer to these data items assigns high membership values to these clearly outlier data items. In the proposed method, however, the two clusters treat these data items similarly and both assign low membership values to them. In other words, in the results of the proposed method, a cluster does not "care" that it is closer to a particular data item than other clusters if it is already "too far" from it. This is a direct result of the outlier recognition mechanism embedded in the developed method.

Moreover, it is informative to compare the geometry of the membership functions to the first cluster from the left in the outputs of the two algorithms. In fact, close assessment of the output of FCM, shown in bottom row of Figure 4.2-(a), reveals that the peak of the membership function is pulled to the left, away from the peak of the histogram. The same cluster in the output of the proposed method, however, as seen in the bottom row of Figure 4.2-(b), exhibits that the peaks of the two curves line up vertically. In other words, the cluster produced by the proposed method is better aligned with the distribution of the data. This phenomenon is a result of the outlier removal mechanism contained in the proposed method.

In the two cases reported here, the FCM and the proposed method require 12 milliseconds

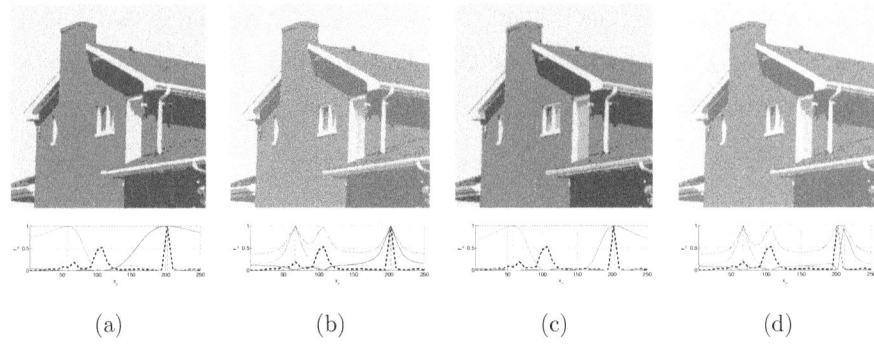

(a)         (b)         (c)         (d)

Figure 4.3: Results of applying FCM and the developed method for grayscale image multi-level thresholding on the Standard image *House*, shown in Figure 4.1, for $C = 3$ and $C = 4$. (a) FCM, $C = 3$. (b) Proposed method, $C = 3$. (c) FCM, $C = 4$. (d) Proposed method, $C = 4$.

and 17 milliseconds to conclude, respectively. The input image used in this experiment is at the resolution of $512 \times 512$ pixels.

Figure 4.3 shows additional results produced by FCM as well as the proposed method for the same input image. Here, two values of $C = 3$ and $C = 4$ are utilized. Similar to the observations made in the analysis of Figure 4.2, we notice wider clusters in the outputs of FCM compared to the corresponding ones generated by the proposed method. Moreover, we observe that the membership curves generated by FCM remain high when no other competing clusters are present, even as data items move away from the clusters. This effect is, for example, visible to the left and to the right of the membership curves in the bottom row of Figures 4.3-(a) and -(c).

Close assessment of Figures 4.3-(c) and -(d) shows another important difference between the results of FCM and the proposed method. In fact, when FCM is requested to generate four clusters, it places one around the value of $x_n = 150$, although this section of the histogram is fairly vacant. In contrast, in the proposed method, clusters 3 and 4 divide a peak in the histogram amongst each other.

### 4.2.2 2D Euclidean Clustering

Figure 4.4 presents the results of applying FCM and the developed method on a set of input data items. In this figure, the sizes of the data items denote the corresponding values of $w_n$ and the shades of gray demonstrate $p_n$.

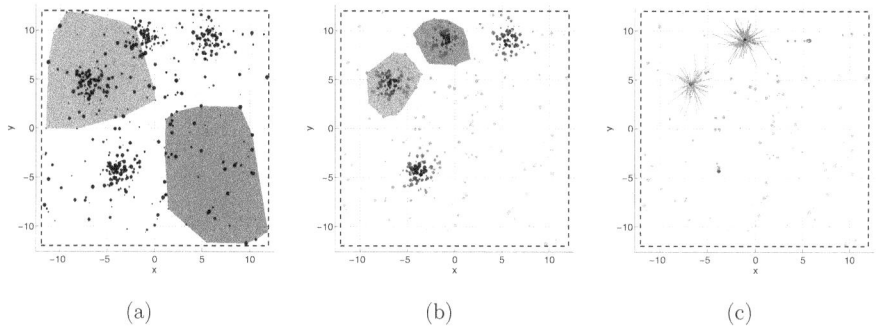

(a)                (b)                (c)

Figure 4.4: Results of applying FCM and the developed method for 2D Euclidean clustering. Size of the data items denote their weight and their shade of gray indicates the probability of being an inlier. (a) FCM. (b) Proposed method. (c) Data items after they have responded to the clusters.

In Figure 4.4 the colored polygons identify the convex hull of the clusters produced by the two algorithms. In other words, any data item outside these polygons is considered by the corresponding algorithm to be an outlier. As expected, darker data items are contained within the polygons. Moreover, FCM includes every data item in the union of the polygons. Hence, we argue that, the clusters produced by the proposed method are more specific and the data items which are far from the clusters are not forced upon the cluster closest to them by the developed method. Similar to the observations made in Section 4.2.1, this difference between the results of the two algorithms is more visible in the boundaries of the data, where clusters are not faced with competition from other clusters.

Close assessment of Figure 4.4-(a) also shows that the two clusters on the positive side of the y-axis have extended their respective sets into a collection of data items which happen to exist between them. These data items are in fact assigned as a separate cluster by the proposed method, as seen in Figure 4.4-(b). In FCM, however, the forces of the two clusters have made this area inhabitable for a third cluster and, thus, the fourth cluster has converged to an area of sparse data items to the bottom right of the domain. This phenomenon is unwanted because of two reasons, i.e. a compact set of data items is ignored and, additionally, a cluster is spotted incorrectly.

In the two cases shown in Figure 4.4, FCM and the proposed method converge after 12 milliseconds and 26 milliseconds, respectively. The set of data items utilized in this experiment contains 591 samples.

34

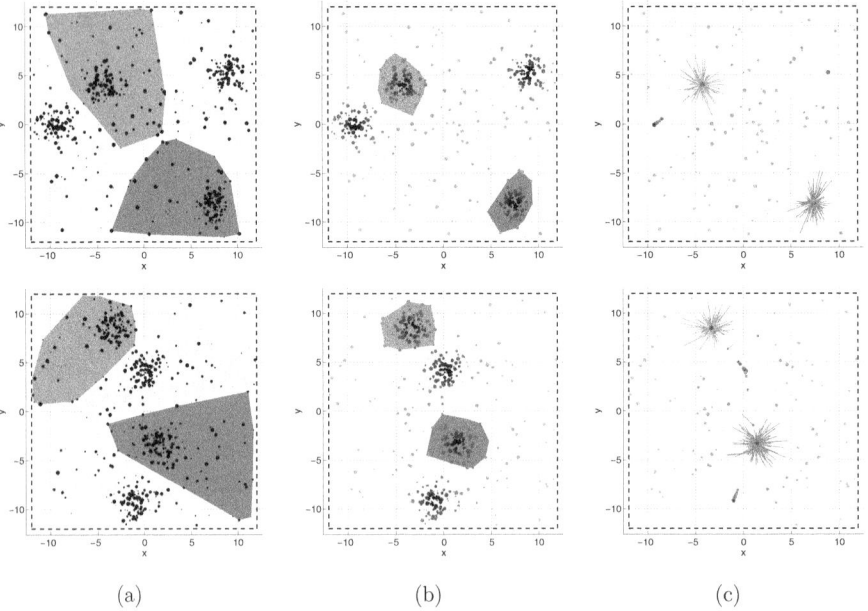

(a)     (b)     (c)

Figure 4.5: Results of applying FCM and the developed method for 2D Euclidean clustering. (a) FCM. (b) Proposed method. (c) Data items after they have responded to the clusters.

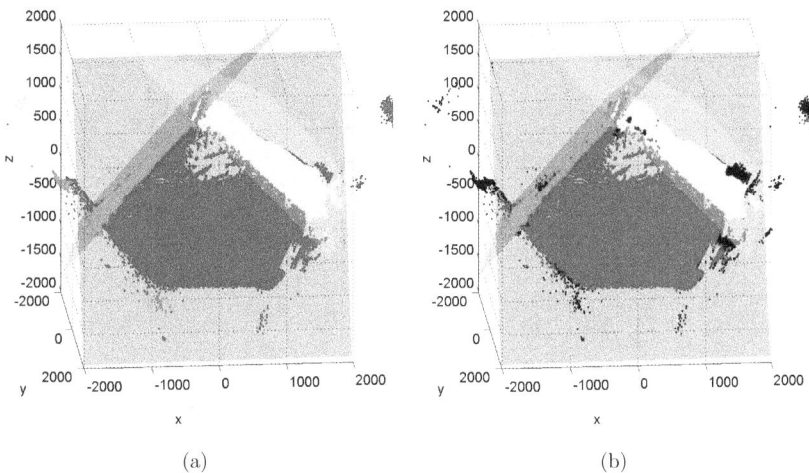

Figure 4.6: Results of applying FCM and the developed method for plane finding in range data. (a) FCM. (b) Proposed method.

Figure 4.5 shows additional comparative results for FCM and the proposed method. Here, again, we observe that FCM clusters inflate in order to cover the entire set of data items. The clusters generated by the proposed method, however, are more specific and reject outliers.

### 4.2.3 Plane Finding in Range Data

Figure 4.6 shows the results of applying FCM and the proposed method for finding planes in range data. This set of data items corresponds to the corner of a room, where two walls and the floor are visible. Here, each translucent plane represents one of the clusters discovered in the data. The data items in these results are painted according to the corresponding values of $c_n$. Hence, as seen in Figure 4.6-(b), there exist data items which are painted in black in the output of the proposed method. These data items are designated as outliers by the proposed method. This result shows similar performance by FCM and the developed method.

While FCM converges in $1,421$ milliseconds, the proposed requires $9,243$ milliseconds to converge. This set contains $172,240$ data items.

While FCM results are comparable to the outputs of the proposed method for the sample shown in Figure 4.6, more careful comparison reveals significant differences between the two algorithms

36

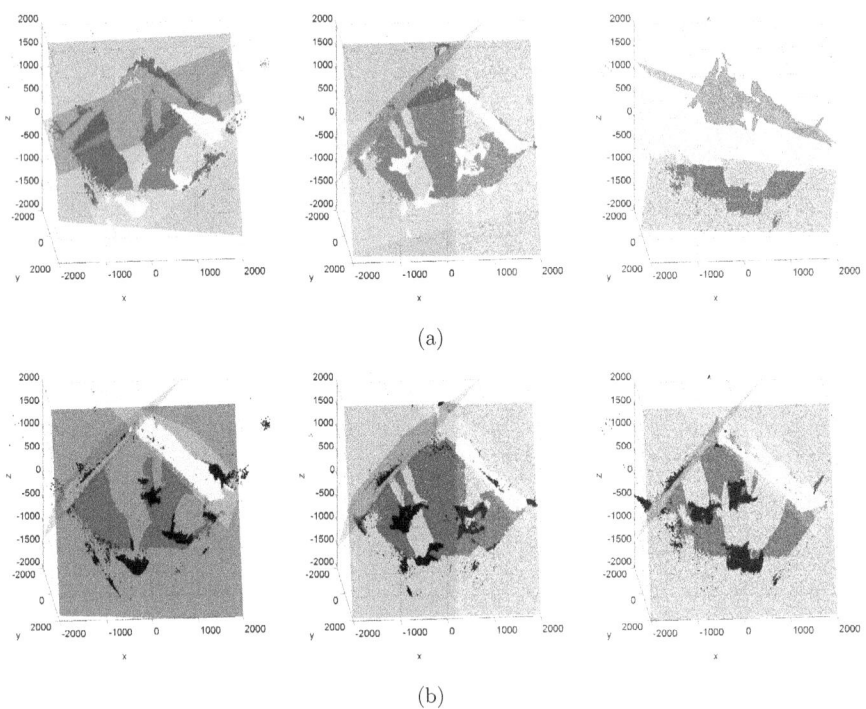

Figure 4.7: Results of applying FCM and the developed method for plane finding in range data. (a) FCM. (b) Proposed method.

in the presence of outliers.

Figure 4.7 shows three additional sets of data items processed by both FCM and the proposed method. These sets of data items correspond to the same room captured in Figure 4.6, with the addition of three human bodies. As observed in Figure 4.7-(a), in the presence of outliers, i.e. the human bodies which do not belong to the walls and the floor, FCM clusters converge to sets which combine portions of the present planes with sections of the bodies of the people in the scene. The proposed method, however, as seen in Figure 4.7-(b), is capable of converging to the planes present in the sets of data items in every case. Additionally, we observe that data items which belong to the bodies of the people present in the scene are correctly distinguished as outliers, and are thus painted in black.

## 4.3    Responsiveness

The results presented in Section 4.2 indicate that the proposed method outperforms FCM in terms of both finding the clusters in the presence of outliers as well as recognizing the outliers. As stated in the beginning of this section, however, the proposed algorithm goes beyond being an alternative to conventional clustering algorithms. In this section, we review the concept of data responsiveness as encompassed by the proposed method. The results carried here belong to the same experiments which are presented in Section 4.2.

### 4.3.1    Grayscale Image Multi-Level Thresholding

Figure 4.8-(a) shows the relationship between input and output data items for the experiment shown in Figure 4.2. As discussed in Section 4.2.1, FCM is commonly followed by a hard classification stage which, in effect, modifies $x_n$ to $\psi_{c_n}$, i.e. the representative of the cluster to which $x_n$ is assigned. This relationship is seen in the dashed curve in Figure 4.8-(a), where $x_n^\star$ accepts two values of $\psi_1$ and $\psi_2$ and there is a hard threshold between the set of data items which are assigned to each.

The proposed method, however, modifies each value of $x_n$ individually, in order to produce a more optimal fit for the set of clusters discovered in the data. This approach results in the solid curve visible in Figure 4.8-(a), where not only $x_n^\star$ accepts a range of values, but also it is a continuous function of $x_n$. The difference between Maximum Likelihood and the proposed method in this respect can also be seen in Figures 4.2 and 4.3, where FCM outputs are 2-, 3-, or 4-level

38

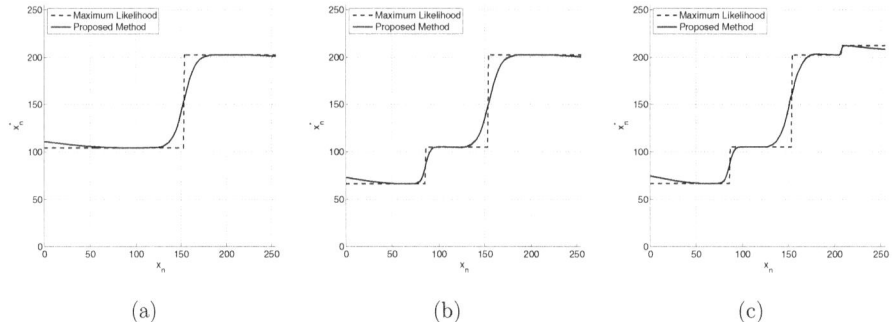

(a)                 (b)                 (c)

Figure 4.8: Data responsiveness for grayscale image multi-level thresholding corresponding to the results shown in Figures 4.2 and Figure 4.3.

images, while the proposed method produces smooth output images.

Figures 4.8-(a) and -(b) show the relationship between Maximum Likelihood and the proposed method for the results presented in Figure 4.3. Here, too, we observe that FCM produces a C-level clustering result while the proposed method generates a smooth output in which $x_n^*$ accepts a range of values.

## 4.3.2   2D Euclidean Clustering

Figure 4.4-(c) shows the process of data responsiveness as it is carried out in the experiment exhibited in Figure 4.4-(b). Here, the data items are points in a 2D plane and they respond to the clusters by moving towards the cluster to which they have a higher level of membership.

The star-shaped structures in Figure 4.4-(c) each represent on cluster. Each ray in these structures denotes one data item as it appears in the input set and the same data item after it responds to the converged clusters. Data items which do not belong to any of the star-shaped structures are outliers, which are left out of the data responsiveness process.

It is important to emphasize that while the data responsiveness process modifies the data items in order to make them more appropriate members of the set of clusters, the data items are not changed into an "ideal" data item. Such a modification is in fact what happens when Maximum-Likelihood style post-processing filters are employed. The data responsiveness component of the proposed method, however, maintains a distribution around the cluster centers in this case. This is the 2D equivalent of the continuous $x_n$-$x_n^*$ relationship curves given in Figure 4.8.

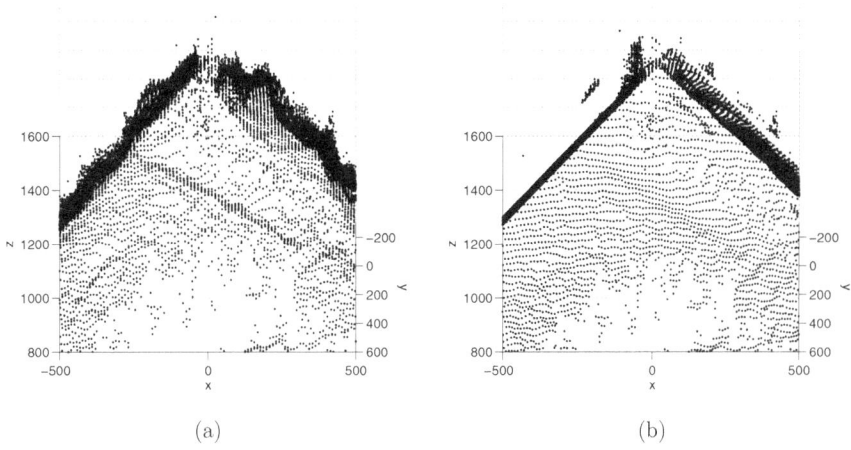

(a)                                                     (b)

Figure 4.9: Examination of the responses of the data items to the converged clusters for the result shown in Figure 4.6. (a) Before. (b) After.

Figure 4.5-(c) shows additional outputs of the data responsiveness process corresponding to the results shown in Figure 4.5-(b). Here, too, we observe that data items converge towards the cluster they have the highest membership to, while the outliers are left intact. Note that the data responsiveness process proposed in this paper does consider the entire set of clusters and the fact that the data items tend towards their most *a posteriori* likely cluster is an inherent quality which emerges from the derivations.

### 4.3.3    Plane Finding in Range Data

Figure 4.9-(a) shows a magnified section of the set of data items utilized in the experiment presented in Figure 4.6. This view corresponds to a corner in the room, but, as clearly visible in Figure 4.9-(a), it is geometrically deformed. This is a result of multi-path and other distortions commonly present in the output of ToF cameras [127]. Figure 4.9-(b) shows the results of data responsiveness for this set of data items. As observed here, deformations along the straight walls are now removed and an orthogonal corner is now clearly distinguishable in the data.

## 4.4 Convergence

As discussed in Section 3.9, this work does not provide proof of convergence for the repeated utilizations of either (3.17) or (3.23). Note that, in both cases, the variable which is to be optimized is involved in the weights, as denoted by $\omega_{nc}^x$ and $\omega_{nc}^\psi$, respectively. We conjecture that the structure of the original Weiszfeld proof, as outlined in [120], may be applicable to this work in certain circumstances as well.

Nevertheless, in this paper, we utilize a practical solution for this important challenge. In fact, in the implementation, we assess the state of the total loss after any call to either $\Xi_\psi(\cdot)$ or $\Xi_x(\cdot)$ and revert the changes if an increase in $\Delta$ is observed. This practice inflates the execution time of the algorithm but, nevertheless, guarantees that the process is guided towards minimizing $\Delta$ in every step.

Additionally, we investigate the impact of the update equations for a large number of sets of data items and assess cases in which an increase in $\Delta$ is registered after a call to either of $\Xi_\psi(\cdot)$ or $\Xi_x(\cdot)$.

Table 4.1: Convergence analysis results.

| Problem Class | # Experiments | $\Xi_\psi(\cdot)$ | | | $\Xi_x(\cdot)$ | | |
| --- | --- | --- | --- | --- | --- | --- | --- |
| | | # Executions | # Failures | Worst Case | # Executions | # Failures | Worst Case |
| Grayscale Image Multi–Level Thresholding | 351 | $150,575$ | $1,121$ $(0.74\%)$ | $2.76\%$ | $90,243,072$ | $0$ $(0.00\%)$ | $0.00\%$ |
| 2D Euclidean Clustering | 200 | $21,196$ | $37$ $(0.17\%)$ | $0.04\%$ | $115,422$ | $0$ $(0.00\%)$ | $0.00\%$ |
| Plane Finding in Range Data | 7 | $1,074$ | $14$ $(1.30\%)$ | $0.00\%$ | $1,239,944$ | $0$ $(0.00\%)$ | $0.00\%$ |

Table 4.2: Convergence analysis results.

| Problem Class | # Experiments | Function | # Executions | # Failures | Worst Case |
|---|---|---|---|---|---|
| Grayscale Image | 351 | $\Xi_\psi(\cdot)$ | $150,575$ | $1,121$ $(0.74\%)$ | $2.76\%$ |
| Multi–Level | | $\Xi_x(\cdot)$ | $90,243,072$ | $0$ $(0.00\%)$ | $0.00\%$ |
| Thresholding | | | | | |
| | | | | | |
| 2D Euclidean | 200 | $\Xi_\psi(\cdot)$ | $21,196$ | $37$ $(0.17\%)$ | $0.04\%$ |
| Clustering | | $\Xi_x(\cdot)$ | $115,422$ | $0$ $(0.00\%)$ | $0.00\%$ |
| | | | | | |
| Plane Finding | 7 | $\Xi_\psi(\cdot)$ | $1,074$ | $14$ $(1.30\%)$ | $0.00\%$ |
| in Range Data | | $\Xi_x(\cdot)$ | $1,239,944$ | $0$ $(0.00\%)$ | $0.00\%$ |

Table 4.2 presents the results of executing the developed method on multiple sets of data items for each one of the three problem classes investigated in this section. In this table, the second column carries the number of experiments for each problem class. Here, a *failure* in an experiment is one instance in which the execution of either $\Xi_\psi(\cdot)$ or $\Xi_x(\cdot)$ results in a new value of $\Delta^\star$ which is larger than the value of $\Delta$ before the update occurs. Hence, any disadvantageous attempt to modify either $\psi_c$ or $x_n$ is registered as one failure. When such a failure occurs, we note the value of $\frac{\Delta - \Delta^\star}{\Delta}$, which is negative, and define the worst case as the magnitude of the smallest such value in the entire set of calls to the respective function.

As seen in Table 4.2, Among the millions of calls made to $\Xi_x(\cdot)$, in the context of three different problem classes, with different values of $C$, no case of failure is registered. In the case of $\Xi_\psi(\cdot)$, however, in a minority of cases, which is less than two in a hundred, a failure is registered, which in the worst case triggers an inflation by less than three percent. Note that, additionally, the fact that the total loss increases in one step is not an indication that the algorithm is to diverge.

We emphasize that this analysis does not provide a formal proof of convergence. It merely suggests that conditions may exist under which for some distance functions, the repeated calls to $\Xi_\psi(\cdot)$ and $\Xi_x(\cdot)$, in the context of the algorithm developed in this paper, may lead to convergence.

# Chapter 5

# Conclusions

In this paper, we analyzed the generic problem of data clustering in a fuzzy class-independent context. We demonstrated that Bayesian inference can be utilized in order to assess the loss for a particular clustering solution. Through this process, we provided an alternative to the available approaches which propose model parameters, objective functions, and constraints based on intuition or heuristics. In fact, we utilized an explicitly derivation-based methodology and showed that the Alternating Optimization framework, which is dominantly used in the literature, can be replaced by an iterative process. We carried experimental results and exhibited the comparative performance of the developed algorithm in terms of recognizing the clusters present in the data as well as rejecting outliers. We also demonstrated that the novel cost function developed in this paper can be utilized in order to allow for data responsiveness.

In fact, as stated in the title, this paper proposes the concept of data responsiveness. In contrast to the data models commonly and dominantly used in the literature, in which the data is merely the subject of clustering, in this work, we exhibited that the data items can respond to the clusters and can effectively reduce the total loss. We utilized sample experimental results in order to exhibit the benefits of data responsiveness in different contexts.

This work utilizes two minimization tasks for which we do not have proof of convergence. Thorough review of this aspect of the developed method is what we intend to follow in the continuation of this research. Nevertheless, we utilized a confirmation stage, through which the outputs of the update equations involved in the proposed model are rejected if they do not contribute to the minimization of the loss induced by the resulting clustering solution. We also tallied the instances in which such a rejection is required and found out that in less than 2% of the cases, the iterative

43

process developed in this paper may lead to an increase in the total loss by less than $3\%$ in one step.

# Acknowledgments

The author wishes to thank the management of Epson Edge for their help and support during the course of this research. The author wishes to thank *Professor James C. Bezdek* and *Professor Witold Pedrycz* for their mentorship. We wish to thank Mahsa Pezeshki and Andrei Rotenstein for proofreading this manuscript. The idea for this research was conceived while camping at the Bruce Peninsula National Park, Canada, with a group of friends.

# Bibliography

[1] R. Duda, P. Hart, Pattern Classification and Scene Analysis, Wiley, New York, 1973.

[2] R. Gray, Y. Linde, Vector quantizers and predictive quantizers for Gauss-Markov sources, IEEE Transactions on Communications 30 (2) (1982) 381–389.

[3] G. H. Ball, D. J. Hall, A clustering technique for summarizing multivariate data, Behavioral Science 12 (2) (1967) 153–155.

[4] T. Kanungo, D. M. Mount, N. S. Netanyahu, C. D. Piatko, R. Silverman, A. Y. Wu, An efficient k-means clustering algorithm: Analysis and implementation, IEEE Transactions on Pattern Analysis and Machine Intelligence 24 (7) (2002) 881–892.

[5] L. A. Zadeh, Fuzzy sets, Information Control 8 (1965) 338–353.

[6] M.-S. Yang, A survey of fuzzy clustering, Mathematical and Computer Modelling 18 (11) (1993) 1–16.

[7] A. Baraldi, P. Blonda, A survey of fuzzy clustering algorithms for pattern recognition. I & II, IEEE Transactions on Systems, Man, and Cybernetics, Part B: Cybernetics 29 (6) (1999) 778–801.

[8] H. Cheng, J.-R. Chen, J. Li, Threshold selection based on fuzzy c-partition entropy approach, Pattern Recognition 31 (7) (1998) 857–870.

[9] J. C. Bezdek, Pattern Recognition with Fuzzy Objective Function Algorithms, Plenum Press, New York, 1981.

[10] J. C. Bezdek, J. M. Keller, R. Krishnapuram, N. R. Pal, Fuzzy Models and Algorithms for Pattern Recognition and Image Processing, Kluwer Academic Publishers, Boston, 1999.

[11] E. H. Ruspini, A new approach to clustering, Information & Control 15 (1) (1969) 22–32.

[12] J. C. Dunn, A fuzzy relative of the ISODATA process and its use in detecting compact well-separated clusters, Journal of Cybernetics 3 (3) (1973) 32–57.

[13] J. M. Leski, Generalized weighted conditional fuzzy clustering, IEEE Transactions on Fuzzy Systems 11 (6) (2003) 709–715.

[14] J. Yu, Q. Cheng, H. Huang, Analysis of the weighting exponent in the FCM, IEEE Transactions on Systems, Man, and Cybernetics, Part B: Cybernetics 34 (1) (2004) 634–639.

[15] M. Trivedi, J. C. Bezdek, Low-level segmentation of aerial images with fuzzy clustering, IEEE Transactions on Systems, Man, and Cybernetics 16 (4) (1986) 589–598.

[16] H. Frigui, R. Krishnapuram, A robust algorithm for automatic extraction of an unknown number of clusters from noisy data, Pattern Recognition Letters 17 (12) (1996) 1223–1232.

[17] F. Klawonn, R. Kruse, H. Timm, Fuzzy shell cluster analysis, in: G. della Riccia, H. Lenz, R. Kruse (Eds.), Learning, networks and statistics, Springer, 1997, pp. 105–120.

[18] J. C. Bezdek, A physical interpretation of fuzzy ISODATA, IEEE Transactions on Systems, Man and Cybernetics SMC-6 (5) (1976) 387–389.

[19] N. R. Pal, J. C. Bezdek, On cluster validity for the fuzzy C-means model, IEEE Transactions on Fuzzy Systems 3 (3) (1995) 370–379.

[20] C. Borgelt, Objective functions for fuzzy clustering, in: C. Moewes, A. Nurnberger (Eds.), Computational Intelligence in Intelligent Data Analysis, Vol. 445 of Studies in Computational Intelligence, Springer Berlin Heidelberg, 2013, pp. 3–16.

[21] K. Zhou, C. Fu, S. L. Yang, Fuzziness parameter selection in fuzzy c-means: The perspective of cluster validation, Science China Information Sciences 57 (11) (2014) 1–8.

[22] M. Lichman, UCI machine learning repository (2013).
URL http://archive.ics.uci.edu/ml

[23] J. C. Bezdek, N. R. Pal, Some new indexes of cluster validity, IEEE Transactions on Systems, Man, and Cybernetics, Part B: Cybernetics 28 (3) (1998) 301–315.

[24] I. Sledge, J. C. Bezdek, T. C. Havens, J. M. Keller, Relational generalizations of cluster validity indices, IEEE Transactions on Fuzzy Systems 18 (4) (2010) 771–786.

[25] I. Ozkan, I. Turksen, Upper and lower values for the level of fuzziness in FCM, in: P. P. Wang, D. Ruan, E. E. Kerre (Eds.), Fuzzy Logic, Vol. 215 of Studies in Fuzziness and Soft Computing, Springer Berlin Heidelberg, 2007, pp. 99–112.

[26] K.-L. Wu, Analysis of parameter selections for fuzzy c-means, Pattern Recognition 45 (1) (2012) 407–415.

[27] W. Pedrycz, H. Izakian, Cluster-centric fuzzy modeling, IEEE Transactions on Fuzzy Systems 22 (6) (2014) 1585–1597.

[28] P. J. Rousseeuw, E. Trauwaert, L. Kaufman, Fuzzy clustering with high contrast, Journal of Computational and Applied Mathematics 64 (1-2) (1995) 81–90.

[29] F. Klawonn, F. Hoppner, What is fuzzy about fuzzy clustering? Understanding and improving the concept of the fuzzifier, in: M. R. Berthold, H.-J. Lenz, E. Bradley, R. Kruse, C. Borgelt (Eds.), Advances in Intelligent Data Analysis V, Vol. 2810 of Lecture Notes in Computer Science, Springer Berlin Heidelberg, 2003, pp. 254–264.

[30] F. Klawonn, Fuzzy clustering: Insights and a new approach, Mathware and soft computing 11 (2004) 125–142.

[31] W. Pedrycz, Conditional fuzzy C-means, Pattern Recognition Letters 17 (6) (1996) 625–631.

[32] W. Pedrycz, Fuzzy set technology in knowledge discovery, Fuzzy Sets and Systems 98 (3) (1998) 279–290.

[33] W. Pedrycz, Conditional fuzzy clustering in the design of radial basis function neural networks, IEEE Transactions on Neural Networks 9 (4) (1998) 601–612.

[34] K. K. Chintalapudi, M. Kam, The credibilistic fuzzy C-means clustering algorithm, in: IEEE International Conference on Systems, Man, and Cybernetics (SMC 1998), Vol. 2, 1998, pp. 2034–2039.

[35] K. K. Chintalapudi, M. Kam, A noise-resistant fuzzy C means algorithm for clustering, in: Proceedings of IEEE World Congress on Computational Intelligence, Vol. 2, 1998, pp. 1458–1463.

[36] M.-S. Yang, K.-L. Wu, Unsupervised possibilistic clustering, Pattern Recognition 39 (1) (2006) 5–21.

[37] J. Noordam, W. van den Broek, L. Buydens, Multivariate image segmentation with cluster size insensitive fuzzy C-means, Chemometrics and Intelligent Laboratory Systems 64 (1) (2002) 65–78.

[38] J. M. Jolion, P. Meer, S. Bataouche, Robust clustering with applications in computer vision, IEEE Transactions on Pattern Analysis and Machine Intelligence 13 (8) (1991) 791–802.

[39] P. D'Urso, Fuzzy clustering of fuzzy data, in: J. V. de Oliveira, W. Pedrycz (Eds.), Advances in Fuzzy Clustering and its Applications, Wiley, England, 2007, pp. 155–192.

[40] A. Abadpour, A. S. Alfa, J. Diamond, Video-on-demand network design and maintenance using fuzzy optimization, IEEE Transactions on Systems, Man, and Cybernetics, Part B: Cybernetics 38 (2) (2008) 404–420.

[41] L. Szilagyi, Z. Benyo, S. Szilagyi, H. S. Adam, MR brain image segmentation using an enhanced fuzzy C-means algorithm, in: Proceedings of the 25th Annual International Conference of the IEEE Engineering in Medicine and Biology Society (EMBS 2003), Vol. 1, 2003, pp. 724–726.

[42] W. Cai, S. Chen, D. Zhang, Fast and robust fuzzy C-means clustering algorithms incorporating local information for image segmentation, Pattern Recognition 40 (3) (2007) 825–838.

[43] R. J. Hathaway, Y. Hu, Density-weighted fuzzy C-means clustering, IEEE Transactions on Fuzzy Systems 17 (1) (2009) 243–252.

[44] Y. Yang, Image segmentation based on fuzzy clustering with neighborhood information, Optica Applicata 39 (1) (2009) 135–147.

[45] R. Nock, F. Nielsen, On weighting clustering, IEEE Transactions on Pattern Analysis and Machine Intelligence 28 (8) (2006) 1223–1235.

[46] J.-L. Chen, J.-H. Wang, A new robust clustering algorithm-density-weighted fuzzy C-means, in: Proceedings of IEEE International Conference on Systems, Man, and Cybernetics (SMC 1999), Vol. 3, 1999, pp. 90–94.

[47] A. H. Hadjahmadi, M. M. Homayounpour, S. M. Ahadi, Bilateral weighted fuzzy C-means clustering, Iranian Journal of Electrical & Electronic Engineering 8 (2012) 108–121.

[48] A. M. Bensaid, L. O. Hall, J. C. Bezdek, L. P. Clarke, Partially supervised clustering for image segmentation, Pattern Recognition 29 (5) (1996) 859–871.

[49] C.-H. Li, W.-C. Huang, B.-C. Kuo, C.-C. Hung, A novel fuzzy weighted C-means method for image classification, International Journal of Fuzzy Systems 10 (3) (2008) 168–173.

[50] C.-C. Hung, S. Kulkarni, B.-C. Kuo, A new weighted fuzzy C-means clustering algorithm for remotely sensed image classification, IEEE Journal of Selected Topics in Signal Processing 5 (3) (2011) 543–553.

[51] R. Kruse, C. Doring, M.-J. Lesot, Fundamentals of fuzzy clustering, in: J. V. de Oliveira, W. Pedrycz (Eds.), Advances in Fuzzy Clustering and its Applications, Wiley, England, 2007, pp. 3–29.

[52] R. Yager, D. Filev, Approximate clustering via the mountain method, IEEE Transactions on Systems, Man and Cybernetics 24 (8) (1994) 1279–1284.

[53] G. Beni, X. Liu, A least biased fuzzy clustering method, IEEE Transactions on Pattern Analysis and Machine Intelligence 16 (9) (1994) 954–960.

[54] J. M. Leski, Fuzzy c-varieties/elliptotypes clustering in reproducing kernel Hilbert space, Fuzzy Sets and Systems 141 (2) (2004) 259–280.

[55] A. K. Jain, R. C. Dubes, Algorithms for Clustering Data, Prentice-Hall, 1981.

[56] D.-M. Tsai, C.-C. Lin, Fuzzy C-means based clustering for linearly and nonlinearly separable data, Pattern Recognition 44 (8) (2011) 1750–1760.

[57] K.-L. Wu, M.-S. Yang, Alternative C-means clustering algorithms, Pattern Recognition 35 (10) (2002) 2267–2278.

[58] L. Chen, C. Chen, M. Lu, A multiple-kernel fuzzy C-means algorithm for image segmentation, IEEE Transactions on Systems, Man, and Cybernetics, Part B: Cybernetics 41 (5) (2011) 1263–1274.

[59] S. Chen, D. Zhang, Robust image segmentation using FCM with spatial constraints based on new kernel-induced distance measure, IEEE Transactions on Systems, Man, and Cybernetics, Part B: Cybernetics 34 (4) (2004) 1907–1916.

[60] K. Honda, N. Sugiura, H. Ichihashi, Fuzzy PCA-guided robust k-means clustering, IEEE Transactions on Fuzzy Systems 18 (1) (2010) 67–79.

[61] H. Zha, C. Ding, M. Gu, X. He, H. Simon, Spectral relaxation for K-means clustering, in: Proceedings of Advances in Neural Information Processing Systems, 2002, pp. 1057–1064.

[62] R. J. Hathaway, J. W. Davenport, J. C. Bezdek, Relational duals of the C-means clustering algorithms, Pattern Recognition 22 (2) (1989) 205–212.

[63] R. J. Hathaway, J. C. Bezdek, NERF C-means: Non-Euclidean relational fuzzy clustering, Pattern Recognition 27 (3) (1994) 429–437.

[64] S. Nascimento, B. Mirkin, F. Moura-Pires, Multiple prototype model for fuzzy clustering, in: D. J. Hand, J. N. Kok, M. R. Berthold (Eds.), Advances in Intelligent Data Analysis, Vol. 1642 of Lecture Notes in Computer Science, Springer Berlin Heidelberg, 1999, pp. 269–279.

[65] L. Fu, E. Medico, FLAME, A novel fuzzy clustering method for the analysis of DNA microarray data, BMC Bioinformatics 8 (3).

[66] T. Hastie, R. Tibshirani, J. Friedman, The Elements of Statistical Learning, Springer, New York, 2009.

[67] J. C. Bezdek, R. J. Hathaway, VAT: A tool for visual assessment of (cluster) tendency, in: Proceedings of the 2002 International Joint Conference on Neural Networks (IJCNN 2002), Vol. 3, 2002, pp. 2225–2230.

[68] L. Wang, U. T. Nguyen, J. C. Bezdek, C. A. Leckie, K. Ramamohanarao, iVAT and aVAT: Enhanced visual analysis for cluster tendency assessment, in: M. J. Zaki, J. X. Yu, B. Ravindran, V. Pudi (Eds.), Advances in Knowledge Discovery and Data Mining, Vol. 6118 of Lecture Notes in Computer Science, Springer Berlin Heidelberg, 2010, pp. 16–27.

[69] T. C. Havens, J. C. Bezdek, An efficient formulation of the improved visual assessment of cluster tendency (iVAT) algorithm, IEEE Transactions on Knowledge and Data Engineering 24 (5) (2012) 813–822.

[70] K. Jajuga, $L_1$-norm based fuzzy clustering, Fuzzy Sets and Systems 39 (1) (1991) 43–50.

[71] L. Bobrowski, J. C. Bezdek, C-means clustering with the $\ell_1$ and $\ell_\infty$ norms, IEEE Transactions on Systems, Man, and Cybernetics 21 (3) (1991) 545–554.

[72] R. J. Hathaway, J. C. Bezdek, Y. Hu, Generalized fuzzy C-means clustering strategies using $L_p$ norm distances, IEEE Transactions on Fuzzy Systems 8 (5) (2000) 576–582.

[73] N. B. Karayiannisa, M. M. Randolph-Gips, Non-Euclidean C-means clustering algorithms, Intelligent Data Analysis 7 (2003) 405–425.

[74] D. E. Gustafson, W. C. Kessel, Fuzzy clustering with a fuzzy covariance matrix, in: IEEE Conference on Decision and Control including the 17th Symposium on Adaptive Processes, Vol. 17, San Diego, CA, 1979, pp. 761–766.

[75] I. Gath, A. Geva, Unsupervised optimal fuzzy clustering, IEEE Transaction on Pattern Analysis Machine Intelligence 11 (7) (1989) 773–781.

[76] J. C. Bezdek, C. Coray, R. Gunderson, J. Watson, Detection and characterization of cluster substructure I. linear structure: Fuzzy c-lines, SIAM Journal on Applied Mathematics 40 (2) (1981) 339–357.

[77] J. C. Bezdek, C. Coray, R. Gunderson, J. Watson, Detection and characterization of cluster substructure II. Fuzzy c-varieties and convex combinations thereof, SIAM Journal on Applied Mathematics 40 (2) (1981) 358–372.

[78] I. Anderson, J. C. Bezdek, R. Dave, Polygonal shape description of plane boundaries, Systems science and science 1 (1982) 295–301.

[79] R. W. Gunderson, An adaptive FCV clustering algorithm, International Journal of Man-Machine Studies 19 (1) (1983) 97–104.

[80] R. Krishnapuram, H. Frigui, O. Nasraoui, Quadratic shell clustering algorithms and the detection of second-degree curves, Pattern Recognition Letters 14 (1993) 545–552.

[81] R. Krishnapuram, H. Frigui, O. Nasraoui, Fuzzy and possibilistic shell clustering algorithms and their application to boundary detection and surface approximation - part I, IEEE Transaction on Fuzzy Systems 3 (1) (1995) 29–43.

[82] R. Krishnapuram, H. Frigui, O. Nasraoui, Fuzzy and possibilistic shell clustering algorithms and their application to boundary detection and surface approximation - Part II, IEEE Transaction on Fuzzy Systems 3 (1) (1995) 44–60.

[83] H. Frigui, R. Krishnapuram, A comparison of fuzzy shell-clustering methods for the detection of ellipses, IEEE Transactions on Fuzzy Systems 4 (2) (1996) 193–199.

[84] R. N. Dave, R. Krishnapuram, Robust clustering methods: A unified view, IEEE Transactions on Fuzzy Systems 5 (2) (1997) 270–293.

[85] J. Leski, Towards a robust fuzzy clustering, Fuzzy Sets and Systems 137 (2) (2003) 215–233.

[86] P. D'Urso, L. D. Giovanni, Robust clustering of imprecise data, Chemometrics and Intelligent Laboratory Systems 136 (2014) 58–80.

[87] R. N. Dave, Characterization and detection of noise in clustering, Pattern Recognition Letters 12 (11) (1991) 657–664.

[88] Y. Ohashi, Fuzzy clustering and robust estimation, Presented at the 9th SAS Users Group International (SUGI) Meeting at Hollywood Beach, Florida. (1984).

[89] R. N. Dave, Robust fuzzy clustering algorithms, in: Second IEEE International Conference on Fuzzy Systems, Vol. 2, 1993, pp. 1281–1286.

[90] R. Krishnapuram, J. M. Keller, A possibilistic approach to clustering, IEEE Transactions on Fuzzy Systems 1 (2) (1993) 98–110.

[91] N. R. Pal, K. Pal, J. M. Keller, J. C. Bezdek, A possibilistic fuzzy c-means clustering algorithm, IEEE Transactions on Fuzzy Systems 13 (4) (2005) 517–530.

[92] M. Barni, V. Cappellini, A. Mecocci, Comments on "A possibilistic approach to clustering", IEEE Transactions on Fuzzy Systems 4 (3) (1996) 393–396.

[93] H. Timm, C. Borgelt, C. Doring, R. Kruse, An extension to possibilistic fuzzy cluster analysis, Fuzzy Sets and Systems 147 (1) (2004) 3–16.

[94] R. Dave, S. Sen, On generalising the noise clustering algorithms, in: Proceedings of the 7th IFSA World Congress (IFSA 1997), 1997, pp. 205–210.

[95] N. R. Pal, K. Pal, J. C. Bezdek, A mixed c-means clustering model, in: Proceedings of the Sixth IEEE International Conference on Fuzzy Systems, Vol. 1, 1997, pp. 11–21.

[96] N. R. Pal, K. Pal, J. M. Keller, J. C. Bezdek, A new hybrid C-means clustering model, in: Proceedings of the 2004 IEEE International Conference on Fuzzy Systems, Vol. 1, 2004, pp. 179–184.

[97] X.-Y. Wang, J. M. Garibaldi, Simulated annealing fuzzy clustering in cancer diagnosis, Informatica 29 (1) (2005) 61–70.

[98] A. Keller, Fuzzy clustering with outliers, in: Proceesings of the 19th International Conference of the North American Fuzzy Information Processing Society (NAFIPS 2000), 2000, pp. 143–147.

[99] D.-Q. Zhang, S.-C. Chen, A comment on "Alternative C-means clustering algorithms", Pattern Recognition 37 (2) (2004) 173–174.

[100] R. Krishnapuram, C.-P. Freg, Fitting an unknown number of lines and planes to image data through compatible cluster merging, Pattern Recognition 25 (4) (1992) 385–400.

[101] R. Krishnapuram, O. Nasraoui, H. Frigui, The fuzzy C-spherical shells algorithm: a new approach, IEEE Transactions on Neural Networks 3 (5) (1992) 663–671.

[102] R. N. Dave, T. Fu, Robust shape detection using fuzzy clustering: Practical applications, Fuzzy Sets and Systems 65 (2-3) (1994) 161–185.

[103] H. Frigui, R. Krishnapuram, Clustering by competitive agglomeration, Pattern Recognition 30 (7) (1997) 1109–1119.

[104] C. J. Friedrich, Alfred Weber's Theory of the Location of Industries, Chicago University Press, Chicago, 1929, translated from the original title "Uber den Standort der Industrien" by Alfred Weber.

[105] G. Wesolowski, The Weber problem: History and perspective, Location Science 1 (1993) 5–23.

[106] E. Weiszfeld, Sur le point pour lequel la somme des distances de n points donnes est minimum, Tohoku Mathematical Journal 43 (1937) 355–386.

[107] Z. Drezner, A note on accelerating the Weiszfeld procedure, Location Science 3 (1995) 275–279.

[108] H. Curry, The method of steepest descent for nonlinear minimization problems, Quarterly of Applied Mathematics 2 (1944) 258–261.

[109] H. W. Kuhn, R. E. Kuenne, An efficient algorithm for the numerical solution of the generalized Weber problem in the spatial economics, Journal of Regional Science 4 (1962) 21–33.

[110] J. Brimberg, R. F. Love, Local convergence in a generalized Fermat-Weber problem, Annals of Operations Research 40 (1) (1992) 33–66.

[111] J. Brimberg, R. F. Love, Global convergence of a generalized iterative procedure for the minisum location problem with $\ell_p$ distances, Operations Research 41 (6) (1993) 1153–1163.

[112] J. Brimberg, R. Chen, D. Chen, Accelerating convergence in the Fermat-Weber location problem, Operations Research Letters 22 (4-5) (1998) 151–157.

[113] A. Cohen, Stepsize analysis for descent methods, Journal of Optimization Theory and Applications 33 (2) (1981) 187–205.

[114] M. R. Osborne, Finite Algorithms in Optimization and Data Analysis, John Wiley, New York, 1985.

[115] A. Bjorck, Numerical Methods for Least Squares Problems, Society for Industrial and Applied Mathematics, 1996.

[116] A. N. Tikhonov, V. Y. Arsenin, Solutions of ill posed problems, Mathematics of Computation 32 (144) (1977) 1320–1322.

[117] K. Levenberg, A method for the solution of certain non-linear problems in least squares, Quarterly Journal of Applied Mathmatics II (2) (1944) 164–168.

[118] D. W. Marquardt, An algorithm for least-squares estimation of nonlinear parameters, Journal of the Society for Industrial and Applied Mathematics 11 (2) (1963) 431–441.

[119] J. J. More, The Levenberg-Marquardt algorithm: Implementation and theory, in: G. Watson (Ed.), Numerical Analysis, Vol. 630 of Lecture Notes in Mathematics, Springer Berlin Heidelberg, 1978, pp. 105–116.

[120] A. Beck, S. Sabach, Weiszfelds method: Old and new results, Journal of Optimization Theory and Applications (2014) 1–40.

[121] A. Abadpour, Rederivation of the fuzzypossibilistic clustering objective function through Bayesian inference, Fuzzy Sets and Systems 305 (2016) 29–53.

[122] P. W. Holland, R. E. Welsch, Robust regression using iteratively reweighted least squares, Communication Statistics - Theory and Methods A6 (9) (1977) 813–827.

[123] P. J. Huber, E. Ronchetti, Robust Statistics, Wiley, New York, 2009.

[124] H. Frigui, R. Krishnapuram, A robust competitive clustering algorithm with applications in computer vision, IEEE Transactions on Pattern Analysis and Machine Intelligence 21 (5) (1999) 450–465.

[125] R. Maier, Out-of-core bundle adjustment for 3D workpiece reconstruction, Master's thesis, Technische Universitat Munchen, Germany (2013).

[126] M. A. Fischler, R. C. Bolles, Random Sample Consensus: A paradigm for model fitting with applications to image analysis and automated cartography, Communications of the ACM 24 (6) (1981) 381–395.

[127] S. Foix, G. Alenya, C. Torras, Lock-in Time-of-Flight (ToF) cameras: A survey, IEEE Sensors Journal 11 (9) (2011) 1917–1926.